Predominantly orange with black dashes, spots, etc., Greater Fritillaries, pp. 118–127

Predominantly yellow, Cloudless or Orange-barred Sulphur, pp. 63, 64

Medium

Predominantly white, Whites, pp. 52–57

Predominantly yellow or orange with black borders on dorsal side, Sulphurs, pp. 58–62, 65, 66

Predominantly orange with black dashes, spots, etc., pp. 128–133

Extremely long proboscis, American Snout, p. 111

Jagged, irregular wing edges, Anglewings, Tortoiseshells, and Mourning Cloak, pp. 140–151

Butterflies of Pennsylvania

Butterflies *of* Pennsylvania

a field guide

James L. Monroe
and David M. Wright

University of Pittsburgh Press

A John D. S. and Aida C. Truxall Book

Published by the University of Pittsburgh Press, Pittsburgh, Pa., 15260
Copyright © 2017, University of Pittsburgh Press
Manufactured in Canada
Printed on acid-free paper
10 9 8 7 6 5 4 3 2 1

ISBN 13: 978-0-8229-6455-1
ISBN 10: 0-8229-6455-4

Cataloging-in-Publication data is available from the Library of Congress

Cover photograph by James L. Monroe
Cover design by Joel W. Coggins

When we try to pick out anything by itself, we find it hitched to everything else in the Universe.

—*John Muir*

CONTENTS

ACKNOWLEDGMENTS

We acknowledge our deep gratitude to the following individuals who provided encouragement and inspiring discussions regarding a restructured, updated version of coverage of the butterflies and skippers of Pennsylvania: Tom Allen, Daniel Bogar, Richard Boscoe, Jeffery Belth, John Calhoun, Robert Dirig, Frank Fee, David Iftner, Steve Johnson, Norbert Kondla, Curtis Lehman, Betsy Ray Leppo, Gerald McWilliams, Monica Miller, Bob Moul, Charles Oliver, Paul Opler, Harry Pavulaan, Don Phillips, Eric Quinter, John Rawlins, Jane Ruffin, Dale Schweitzer, Art Shapiro, and Jason Weintraub.

We sincerely thank the following individuals and their respective institutions for providing access to institutional reference collections: Don Azuma, Jon Gelhaus, and Jason Weintraub at the Academy of Natural Sciences of Philadelphia; Jacqueline Miller and Lee Miller at the Allyn Museum of Entomology; David Grimaldi, Courtney Richenbacher, Fred Rindge, and Eric Quinter at the American Museum of Natural History; Jessica Ganyor and James Traniello at the Boston University–Boston Society of Natural History; David Daniell at Butler University; Norman Penny at the California Academy of Sciences; Sonja Teraguchi at the Cleveland Museum of Natural History; Robert Dirig at the Cornell University Insect Collection; Reed Watkins at the Dayton Museum of Natural History; John Hallahan at the Delaware County Institute of Science; David Unander at Eastern College; Nezka Pfeifer at the Everhart Museum; Phil Parillo at the Field Museum of Natural History; John Heppner at the Florida State Collection of Arthropods; Andrew Deans at the Frost Entomological Museum; John Bouseman and Kathleen Zieders at the Illinois Natural History Survey; Linda Badger at the Indiana State Museum; Julian Donahue at the Los Angeles County Museum of Natural History; Mo Nielsen at Michigan State University; Stephan Cover, David Furth, and Phil Perkins at the Museum of Comparative Zoology; Brian Harris and Robert Robbins at the National Museum of Natural History (Smithsonian); Phil Ackery and Kim Goodger at the Natural History Museum–London; David Parris at the New Jersey State Museum; John Michalski at the Newark Museum; Alison Mallin at the North Museum of Nature and Science; Kenneth Mark at the Oakes Museum of Natural History; Eric Metzler at the Ohio State Museum of Biological Diversity; Karl Valley at the Pennsylvania Department of Agriculture; John Quimby at the Pennsylvania Department of Forestry; Barbara Barton and Betsy Ray Leppo at the Pennsylvania Natural Diversity Inventory; Walter Mashaka at the Pennsylvania State Museum; Arvin Provonsha at Purdue University; Michael Feyers at the Reading Museum; Claudia Copley at the Royal British Columbia Museum; Chris Darling at the Royal Ontario Museum; Timothy Casey at Rutgers University; Edward Johnson at the Staten Island Institute of Arts and Sciences; Val Jacoski at the Tioga Point Museum; Art Shapiro and Lynn Kimsey at the University of California–Davis, Bohart Museum; Greg Ballmer and Gordon Pratt at the University of California–Riverside; David Wagner at the University of Connecticut; Dale Bray and Tom Wood at the University of Delaware; Steve Marshall at the University of Guelph; Charles Covell at the University of Louisville; Mark O'Brien at the University of Michigan Museum of Zoology; Andrew Binns at the University of Pennsylvania; Eugene Bolt and Susan Glassman at the Wagner Free Institute of Science; David Hess and Yale Sedman at Western Illinois University; and Larry Gall and Charles Remington at the Yale Peabody Museum of Natural History.

We also offer a special "thank you" to Thomas Emmel and Andrew Warren at the McGuire

Center for Lepidoptera and Biodiversity in Gainesville, Florida, as well as to John Rawlins and Vanessa Verdecia at the Carnegie Museum of Natural History in Pittsburgh, Pennsylvania. The vast majority of the specimens seen in this book were photographed in their invaluable institutions.

Particular thanks are extended to librarian Eileen C. Mathias of the Ewell Sale Stewart Library, Academy of Natural Sciences of Philadelphia, for her invaluable aid in retrieving important literary records of Pennsylvania butterflies. We express our appreciation to Caron O'Neill, Senior Geologic Scientist, Department of Conservation and Natural Resources, Bureau of Topographic and Geologic Survey, in Middletown for permission to use the survey's digital shaded-relief map.

We extend our deep, genuine thanks to the following individuals who provided us with substantial numbers of specimens, photographs, notes, observations, and/or reports: Tom Allen, David Amadio, Don Azuma, Jesse Babonis, Timothy Baird, Nancy Baker, Ethan Bancroft, Barbara Barton, George Bercik, Charles Bier, Chris Blazo, Daniel Bogar, Chris Bohinski, Rick Borchelt, Mary Anne Borge, Richard Boscoe, Bob Bouman, Rory Bower, Mark Bowers, Stephen Boyce, Mark Bresler, Cathy Brouse, Sara Jane Brown, Suzanne Butcher, Robert Byers, Marvin Byler, John Calhoun, Blake Campbell, Gary Campbell, Karen Campbell, Roger Carpenter, Roxanne Carstetter, Paul Cavalconte, Nathan Charnock, Steve Collins, Charles Conaway, Ben Coulter, Ethan Cowles, Marcy Cunkelman, Harry Darrow, Bob Davidson, Lynn Davidson, Link Davis, Sara Davison, Jonathan DeBalko, Paul Dennehy, Nancy Dennis, Robert Dirig, Bruce Dixon, Misty Doy, Michael Drake, Jay Drasher, Lori Dunn, Chris Durden, Victoria Dziadosz, Jim Eckart, Gary Edwards, George Ehle, Robert Ehle, Todd Eiben, Sam Eiben, Ted Enterline, Devich Farbotnik, Frank Fee, Betty Ferster, Clorinda Forte-Katz, Ellery Foutch, James Fowles, Evelyn Fowles, Linda Freedman, Stan Galenty, Larry Gall, Christine Gajewski Gallo, Sam Gano, Bob Gardner, Karl Gardner, Jon Gelhaus, Rick Gillmore, William Gleason, G.S. Glenn, Steven Glynn, Arlene Gmitter, Candy Gonzalez, Laurie Goodrich, Bill Grant, Joseph L. Greco Jr., Alan Gregory, Mike Greiman, Ron Grimm, Bob Grosek, Scott Gross, David Grove, David Guzo, Fred Habegger, Tom Halliwell, Jim Hartman, Shelby Heeter, Kevin Heeter, Clifford Hence III, Harry Henderson, Leonard Hess, Linda Hess, David Hess, Erv Hiller, Jason Horn, William Houtz, Jim Hoyson, Harry Hunt, Corey Husic, David Iftner, Warren Jacobs, Robert Jacobs Jr., Grace Jeschke, Steve Johnson, Rudy Keller, Gregory Keller, Richard Kelly, Will Kerling, Bill Kimmich, Stephen Kloiber, Alexander B. Klots, Chris Knoll, Arlene Koch, Glenn Koppel, Rick Koval, George Krizek, John Kunsman, Alex Lamoreaux, Amy D. Langman, Michael Lanzone, John Laskowski, Nan Lawrence, Thomas Le Blanc, Harold Lebo, Ken Lebo, Jenny Lehman, Clarissa Lehman, Sarah Lehman, Curtis Lehman, Lance Leonhardt, Betsy Ray Leppo, Eleni LeVan, Nate Libal, Larry Lloyd, Allan Loudell, Julie Lundgren, Bob Machesney, Diane Machesney, Sister Mary Franceline Malone, Thomas Manley, Cosmos Mariner, Tom Mason, Bob Mayer, Lawrencine Mazzoli, Tony McBride, Eileen McDonnell, Bruce McNaught, Gerald McWilliams, Win Mergott, Holly Merker, Betsy Mescavage, Rick Mikula, Monica Miller, Carol Snow Milne, Marc Minno, Thomas Moeller, Mark Monroe, Peter Montgomery, Katrina Morris, Bob Moul, Leslie Mundell, Bill Murphy, Naomi Murphy, Ichiro Nakamura, Mark Niessner, Robert Noll, Charles Oliver, Paul Opler, Harry Pavulaan, Tom Pawlesh, Doris Pedding, John Peplinski, Scott Perry, Ann Pettigrew, Autumn Pfeiffer, Don Phillips, Gordon Pratt, John Prescott, Mark Priebe, Nick Pulcinella, John Quimby, Eric Quinter, Becky Rau, Tom Raub, John Rawlins, Sally Ray, Ron

Richael, Theo Rickert, Jim Robbins, Peter Robinson, Ronald Roscioli, Matthew Roth, Jane Ruffin, Carmen Santanasia, Anthony Schoch, Allen Schweinsberg, Dale Schweitzer, Mark Scriber, Bud Sechler, Walter Shaffer, Dana Shaffer, Art Shapiro, Ray Shearer, Clark Shiffer, Jesse Shoemaker, Brad Silfies, Robert Silvick, T. Sims, Jerry Skinner, Mike Slater, Joseph Smaglinski, Jim Smith, Greg Smith, Richard Smith Jr., Ron Smith, Samuel Smith, Robert Snetsinger, Sven-Erik Spichiger, Michael Spingola, Alfred Spoo, Greg Stanko, Jerry Stanley, Alton Sternagle, Andrew Strassman, Kellie Susmann, Mark Swartz, Jennifer Taggart, Chuck Tague, Cheri Tenaglia, Cynthia Tenney, David Trently, Bruce Troy, Aden Troyer, Harvey H. Troyer, Harvey N. Troyer, Jerry Troyer, Mary Troyer, Neil Troyer, Rachel Troyer, Samuel Troyer, Norman Tyson, Gary Tyson, Kathy Tyson, Ray Uhlig, Reno Unger, Karl Valley, Marge Van Tassel, Bill Walbek, Yvonne Walbek, Steve Walter, Will Ward, Billy Weber, Doug Wechsler, Mike Weible, Jason Weintraub , Mervin Wenger, John Wheatley, Robert Whitacre, Patricia Whitacre, Chub Wilcox, Anthony Wilkinson, Tom S. Williams, Peter Woods, Ryan Woolwine, Richard Yahner, David Yeany, David Yeany II, Tracy Yingling-Cutri, Walter Zanol, Ben Ziegler, and Harry Zirlin.

Finally we acknowledge the help, support, and belief in our project by individuals at the University of Pittsburgh Press. In Particular, we thank our editor Ms. Sandra Crooms, who enthusiastically supported us from the beginning of the venture, plus Joel W. Coggns, Amberle Sherman, and Alex Wolfe who all helped bring the project to closure.

The status of the butterflies and skippers of our neighboring states of Ohio, New Jersey, and West Virginia have all been recently studied and published. Pennsylvania's lepidopteran fauna was last reviewed sixty-five years ago. It was therefore apparent to us that an up-to-date appraisal of Pennsylvania's butterflies and skippers was needed. We are optimistic that this field guide will fulfill such a goal, as well establish a benchmark for future studies.

A key objective of this field guide is to provide the best possible information needed to identify the butterflies of Pennsylvania. The following basics are emphasized: (1) what they look like, (2) where they fly, and (3) when they fly. For each species reliably recorded in Pennsylvania, we present high-quality photographs of ventral and dorsal sides of both males and females, plus in many instances commonly occurring forms. The photographs are generally life size or larger. Photographs of many of the small species, such as the hairstreaks and grass skippers, are at a larger scale than found elsewhere. These species are often especially difficult to identify and we feel the larger photographs will make their identification easier. The text portion accompanying each species highlights distinctive physical marks and behavioral traits. Where Pennsylvania butterflies occur is covered by a state map showing the counties where each species has been documented. This map is color-coded to distinguish between historical records (pre-1995) and current records (from the last two decades). A notation of preferred habitat also directs the butterfly enthusiast to where they may be found. Lastly, flight records from over a hundred-year period have been compiled for each species. These data are presented in a bar chart consisting of seven- to eight-day periods from which details of flight time and the number of broods can be extracted.

We have tried to produce a useful field guide where all the information regarding a particular species is generally found on a single page or two facing pages. How helpful the book is, in reality, can only be judged by the reader.

Butterflies of Pennsylvania

Butterflies bring us a moment of joy when they fly freely through fields of flowers or stop to visit our flowery gardens. These magnificent creatures fascinate us with their beauty and amazing metamorphosis from caterpillar to adult butterfly. Our ancestors saw them as symbols of rebirth and reawakening. Today we see them as part of the rich diversity of life and a special gift that we have inherited. Pennsylvania, with its diverse geological and biological features, offers a home for a great variety of butterflies. Of the 800-plus butterfly species recorded in North America north of Mexico, 156 have been recorded in Pennsylvania to date. This diversity makes butterfly watching in Pennsylvania both inspiring and satisfying. In mounting numbers, butterfly enthusiasts are going into the field to observe butterflies in their natural habitat.

Whether we think of ourselves as amateur naturalists, conservationists, or professional biologists, habitat preservation is a prime concern of our generation—we wish to preserve healthy ecosystems and leave a legacy for future generations to enjoy. Butterflies may serve as "quality of life" indicators of their environment and call attention to areas in need of protection. The importance of proper identification of butterflies, recording facets of their life history, and surveying their changing numbers and distribution cannot be overstressed when gathering data for habitat preservation. The study of butterflies never stops; it may consist of a stroll through a garden, a pleasurable day trip, or a lifetime of scholarship. Data from all sources help advance knowledge of our state butterfly fauna.

Evolutionary Origins of Butterflies

Butterflies are arthropods, a group of animals with segmented bodies, jointed legs, and a hard exoskeleton. The most common arthropods are insects and spiders. Insects, including butterflies and moths, are characterized by a three-part body and three pairs of jointed legs.

Butterflies and moths form the order Lepidoptera ("scale wings"), featuring tiny scales that cover the wings and body. Worldwide, there are approximately 18,000 species of butterflies and 180,000 species of moths. The vast majority of them reside in the tropics. The closest living relatives of butterflies and moths are the caddisflies (Trichoptera, "hair wings"). Caddisflies have aquatic larvae that are covered with tiny hairs. The ancestors of the Trichoptera and Lepidoptera split in the Triassic nearly 200 million years ago. The earliest modern lepidopteran families appeared in the Jurassic about 150 million years ago and roamed with the dinosaurs. Soon after the Cretaceous-Tertiary extinction event 65 million years ago, butterflies diversified with flowering plants (angiosperms), their most common larval hosts. Butterfly fossils from this period look surprisingly similar to extant species, indicating that they have been around for a long time in their current form.

Butterfly Anatomy

The adult butterfly fundamentally consists of a body with two pairs of wings, forewings and hindwings. The body contains three sections: (1) head, (2) thorax, and (3) abdomen.

The head serves primarily for feeding and sensory perception. It consists of a hard capsule with two large compound eyes, one on each side. The compound eyes are composed of several hundred individual six-sided simple eyes known as ommatidia. Most butterflies see very well. They can readily detect movement, such as a human walking by or another butterfly intruding in their territory. Sight also is important in finding a potential mate.

Butterfly eyes are able to see colors in the same visible spectrum as ours, plus "colors" in

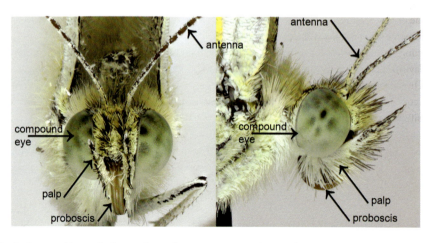

Head of a typical butterfly showing base of antenna, compound eye, palpi, and proboscis. Notice the segmented nature of the antennae and the two halves of the proboscis coupled together. The pattern seen in the compound eye is due to a diffraction of light by hundreds of simple eyes (ommatidia).

the ultraviolet (UV) range. Many plants have structures that are visible only in UV. On top of the head sits a pair of long antennae. Each antenna is composed of multiple segments, with those segments near the tip enlarged and shaped into a club. Numerous tiny sensory receptors cover the surface of the antennae. These detectors work together with compound eyes to help the butterfly find mates, nectar flowers, and potential hostplants. On the underside of the head is a pair of hairy sensory structures called the *palpi*, or *palps*. In most butterflies the palpi curve upward to partially cover the face. Between the palpi sits the proboscis, sometimes referred to as the "tongue." The proboscis is made up of two halves and functions like a flexible drinking straw. When not in use, it is curled up between the palpi. Muscles and vascular channels throughout its length help to uncoil it when called into use. A dense concentration of tiny sensory receptors near the tip helps the butterfly find nectar in flowers.

The thorax is the compact middle section of the body that primarily serves for locomotion. Legs and wings attach here. The thorax consists of three fused segments. The first segment, the prothorax, bears a pair of legs. The middle segment, the mesothorax, bears the middle legs and the first pair of wings (forewings). The last segment, the metathorax, bears the hind legs and second pair of wings (hindwings). Oxygen is supplied to wing and leg muscles via breathing holes, or spiracles, on each side of the thorax. Each leg has several sections. From beginning to end, these sections are the coxa, trochanter, femur, tibia, and tarsus, respectively. The tibia often has rows of little spines with one or two enlarged as spurs. The tarsus also has numerous spines and a claw at the tip. Female butterflies use tarsal spines to investigate plants to determine if they are proper hostplants for larvae. Many butterflies have a small projection of the foretibia called the epiphysis, which they use to clean their antennae.

The abdomen is the long final section of the butterfly body that serves primarily for digestion and reproduction. It is composed of ten segments with the terminal segments modified for mating, egg laying, and excretion. Males have a pair of claspers (also called valvae or valves) on the tip of their abdomen that hold the female abdomen during mating. Females have two separate reproductive openings; one is used for mating (bursa copulatrix) and the other for laying eggs (ovipositor).

On the right is a photo showing the majority of the butterfly leg. The coxa and trochanter are obscured by hairs on the thorax. They are situated close to the thorax and form a movable joint.

clasper

bursa copulatrix

ovipositor

On the left are the male and female final segments of the abdomen. The examples allow one to see the male claspers and the female bursa copulatrix and ovipositor.

The wings are formed from a double

layer of very thin exoskeletal sheets extending from the thorax. Hollow tubular veins lie between the two layers. The pattern of veins in the wings is relatively constant within each butterfly family. The veins provide support and strength for the fragile wings. Thousands of individual scales are attached to the wings and the body; each is formed by a single cell. Most wing scales are broad and flat with rows of them overlapping like shingles. Patterns of color created by scales are important for blending with the environment, startling potential predators, and recognizing the opposite sex. Some scales provide color through pigments (red, brown, yellow, black, and white). Other scales produce structural colors (iridescent blues and greens). These scales have ridges that act as diffraction gratings that break up white light into its component colors. One color is reflected, while other wavelengths are absorbed by melanin embedded in the same scale or an underlying scale. Lastly, scales of many male butterflies are often modified to store scents (pheromones) that are used in courtship.

Specific patterns, colorings, etc., on a particular part of the wing are often used to identify various species. The terminology used to denote various positions on butterfly wings is illustrated below.

Butterflies, Skippers, and Moths

The word *butterfly* is used by some to indicate both "true" butterflies (superfamily Papilionoidea) and skippers (superfamily Hesperioidea). Because of their coloring and size, many beginners may think of skippers as moths. So, how does one differentiate between butterflies, skippers, and moths?

In general, butterflies and skippers are active during the day (diurnal), while most moths are active at night (nocturnal). The antennae of most butterflies are clubbed at the tip; those of moths are either threadlike or feathery (see below). The majority of butterflies have brightly colored wings that are held vertically above their bodies when at rest; the majority of moths have cryptically colored, camouflaged wings held flat or folded tightly over their bodies. There are exceptions to these general statements. In the tropics several day-flying moths have evolved butterflylike characteristics, such as clubbed antennae and colorful wings. In North America, many of the sphingid moths (hawk moths) are day fliers.

Skippers, for the most part, have stouter bodies than butterflies and their colors frequently consist of subdued shades of black, brown, and gray. Skipper antennae differ from true butterflies'; their antennae terminate in a tapering point (often hooked, see below). True butterflies have antennae consisting of a single filament broadening at the end. Skippers have the same with the addition of a hooked portion at the end, known as the apiculus. Moths have a large variety of antenna types, from those of a single filament to those with many feathery branches.

Shown below, from left to right, the top row of five antennae are from representatives of true butterflies: Papilionidae, Pieridae, Lycaenidae, Riodinidae, and Nymphalidae. The lower row of five antennae are from three representatives of Hesperiidae (skippers) and two moths. Note the varying degrees of hooked tips of skipper antennae and the range of shapes of moth antennae.

Scientific Names and Butterfly Classification

Every plant and animal is currently classified in a system of scientific names called binomial nomenclature. A unique two-part name consisting of a genus name and a species name is given to both living and fossil organisms. This scheme was introduced by Swedish biologist Carl Linnaeus in the eighteenth century and it was first utilized for butterflies in 1758.

A species is defined as a population of similar individuals that are known to interbreed or thought to be capable of interbreeding. A genus name is a conceptual notion used to group closely related species. Specialists often disagree as to how many species should be placed in a genus. For the most part, butterfly species in Pennsylvania are well defined and easy to identify. In extraordinary instances, the limits between species can be confusing. For example, the Spring Azure was treated for many years as a single species. But recent investigation has shown that a complex of several species was flying together under one name (see Special Topic: Azure Complex). Butterfly species quite often vary in their appearance in different parts of their range. When a geographically separated population of a single species looks distinctive, a subspecies name can be used to formally identify that population. In many instances the variation between separated populations is continuous and subspecies names are inappropriate.

Above the genus, there is a hierarchy of additional groupings. Genera are contained within subfamilies, subfamilies within families, families within superfamilies, and so on. Diagrams like the one shown below, called tree diagrams, help to visualize the relationships between the various groupings. This diagram shows the current understanding of relationships of the superfamilies, families, and subfamilies of the two superfamilies covered in this book. Items in red indicate that there is a species recorded from Pennsylvania occurring in that group; the number of species recorded from Pennsylvania is shown for each subfamily. Taxonomy, the study of scientific classification, is an ever-changing process. The last word has not been spoken. As better data become available, researchers can apprehend clearer relationships and taxonomic alignments can be revised.

In this book, when a formal scientific name appears, the species or subspecies name is followed by the name of the person who originally proposed the name. If the name was proposed in a different genus than the one now used, the author's name is given in parentheses. Lastly,

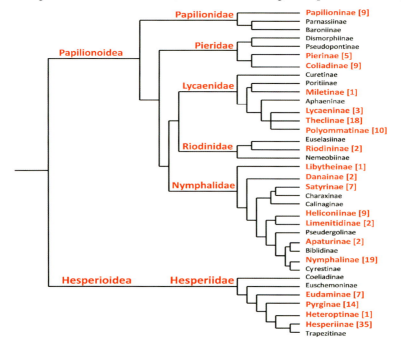

the date of publication in which the name was proposed follows the author's name; sometimes a corrected date is shown in square brackets. In the example of the Mourning Cloak, the scientific name is written *Nymphalis antiopa* (Linnaeus, 1758). In 1758 Linnaeus proposed the name *antiopa* in the genus *Papilio*; over time subsequent revisers assigned the species *antiopa* to the genus *Nymphalis*. Thus the author's name and date are placed in parentheses. The genus and species names are italicized; higher-level groupings such as subfamilies are not italicized.

Butterfly Life Cycle

The complete life cycle of a butterfly proceeds through four stages: egg (ovum), caterpillar (larva), pupa (chrysalis), and adult (imago). This process is called complete metamorphosis. The cycle begins when a female lays an egg. Eggs are glued singly or in clusters on the larval hostplant. The eggs are positioned upright and covered by an outer shell, called the chorion, that is secreted by the female's oviducts. The chorion is often sculpted, giving the egg a distinctive shape and pattern. It is also porous to permit respiration. At the top of the egg are tiny pores called micropyles that allow sperm to enter. After fertilization, eggs usually hatch in three to five days, but in some diapausing species the egg waits till next spring to hatch. The dark head of the developing larva inside the egg is often visible at the top of the egg. The first-stage larva (first instar) chews a hole in the top of the egg and crawls out. In some species the larva consumes the remainder of the egg before starting to eat the hostplant. Like the adult, the butterfly larva has three body sections. The head is a hard capsule with chewing mouthparts, short antennae, and six small eyes on each side of the head. The next section is the thorax. Each of the three thoracic segments bears a pair of segmented legs. The last section of the larva is the abdomen, consisting of ten segments. Fleshy legs (prolegs) occur on abdominal segments 3 through 6, plus on terminal segment 10. Tiny hooks at the tips of the prolegs allow the larva to hold firmly to a twig or leaf. A thin skin (cuticle) covers the thorax and abdomen. The cuticle is often adorned with hairs (setae), simple spines (horns), branching spines (scoli), or fleshy filaments (tubercles) extending from the body. These devices are used principally to deter predators. Some larvae avoid predators by constructing leaf shelters in which to hide during the day, and then resume feeding at night. As the larva grows, the cuticle stretches. However, it cannot stretch indefinitely. The larva must shed its skin (molt) at intervals to allow continual growth. The larva typically passes through four or five instars (some groups have six or more), each one progressively larger. The last instar ultimately molts (pupation) to reveal the pupa. When the last instar stops eating, it discharges food remaining in the gut and may wander a considerable distance to find a place to pupate. Once a site is selected, the larva lays down a mat of silk. At the final molt, the pupa is usually attached to the silk pad by a series of hooks (cremaster) at its rear end; it may also be supported by a silk thread or girdle around its midsection. The pupal stage is often called a chrysalis, a word derived from the Greek word *chrysos*, meaning gold, reflecting the golden-brown color that camouflages many pupae. The pupa occasionally is referred to as a resting stage, because it outwardly gives the appearance of lacking movement. However, when disturbed, pupae can wriggle, thrash, and chirp to ward off predators. Inwardly, tremendous changes take place with the pupa as larval tissues are broken down and resting clusters of cells awaken to create adult structures. If there are no delays in development, the adult may emerge in about ten days. However, the pupal stage may last months or even years in species that diapause. Diapause is an enforced period of physiologic dormancy, usually used to survive winters or prolonged dry spells. As the adult readies to emerge from the pupa, the pupal cuticle becomes transparent. To break out of the pupa, the adult enlarges itself by taking in air through its spiracles. Subsequently, the pupal cuticle splits along lines of weakness and the butterfly crawls out—this is called eclosion. As the butterfly rests, fluid is forced into the wing veins to expand and straighten its wings. Because the pupa lacks an external anal opening, it is unable to excrete metabolic byproducts. The eclosed adult butterfly squirts out this waste fluid (meconium). Within a few hours its wings have hardened and the butterfly is able to fly away.

Ova of American Copper, Banded Hairstreak, and Question Mark.

Larvae of Acadian Hairstreak and Black Swallowtail.

Prepupa and pupa of Common Buckeye, pupa of Monarch, and pupa of Eastern Tiger Swallowtail.

Butterfly Seasons

Most adult butterflies live from a few days to several weeks in nature. During this period their ultimate goal is to locate a mate and lay eggs on an appropriate hostplant. Azures live

Introduction

about a week, sulphurs about a month, and fritillaries about two months. Overwintering anglewings and Monarchs are especially long-lived, lasting eight months or more. Most species have several generations, or broods, per year. A complete brood may require one to three months to complete. Species like the Eastern Tailed-Blue may have four to five broods per year. Many have two broods (satyrs and grass skippers) and a few have just one brood per year (elfins and duskywings). The top seasons for butterflies in Pennsylvania are spring, early summer, and late summer. The first butterflies of the year start to appear in early spring (March through April). Species that overwinter in the pupal stage, such as the Spring Azure, Cabbage White, Tiger Swallowtail, and Juvenal's Duskywing, are the first new butterflies to emerge. Mid-May through June marks the late spring appearance of some lycaenids such as the Banded Hairstreak and Appalachian Azure. These butterflies have only one generation, and the adults are present for only a few weeks up to a month. During early summer (late June through July) the second generation of swallowtails and other butterflies, such as the Baltimore Checkerspot, fritillaries, and grass skippers, emerge. Late summer (late July through mid-September) is the season of large numbers of butterflies. Resident species reach their second or third broods, and many migrating butterflies such as the Cloudless Sulphur, Common Buckeye, and Sachem fly northward by the thousands, while Monarchs are traveling south. The migrants may continue to appear throughout the fall (October and November).

Migration

Some adult butterflies are adept at flying astonishing distances. During the fall, Monarchs fly two thousand miles or more in a spectacular flight from Pennsylvania to their over-wintering grounds in central Mexico. During late summer and early fall, large numbers of Cloudless Sulphurs, Common Buckeyes, Painted Ladies, Variegated Fritillaries, Sachems, Fiery Skippers, Common Checkered-Skippers, and others fly into Pennsylvania from other regions of the United States. Certain dispersal lanes or flyways bring these migrating species into the state in great concentrations, such as the Atlantic Coastal Plain in the eastern part of the state, river corridors in midstate (Susquehanna River valley) and western Pennsylvania (Ohio River valley), and a water boundary along the southern edge of Lake Erie. These migrant species generally cannot withstand freezing temperatures for long and do not survive our winters. Of Pennsylvania's total butterfly fauna, 78% are residents and 22% are migrants. Most migrants arrive from the south (88%), with a smaller percent coming from the west (12%).

Overwintering

Pennsylvania butterflies possess a range of methods to overwinter. Physiologic dormancy (diapause) may occur in any of the four life stages. It may take place during the egg stage (some hairstreaks and coppers), the young larval stage (fritillaries and Common Wood Nymph), when larvae are partially grown (grass skippers) or fully grown (oak-feeding duskywings), during the pupal stage (swallowtails, some pierids and lycaenids), or during the adult stage (some nymphalids). Our butterflies that overwinter, such as the Mourning Cloak, Eastern Comma, and tortoiseshells, seek shelter in hollow trees, wood piles, rock mounds, and even garages during the cold months from November to March. They often venture out on warm days to siphon moisture from damp soil or to imbibe sap escaping from tree wounds. Of Pennsylvania's resident butterfly fauna, the breakdown of over-wintering stages is as follows: egg, 9%; larva, 56%; pupa, 29%; and adult, 6%.

Mating

The primary concern of adult butterflies is mating and passing their genes onto the next generation. Males spend a significant amount of time searching for mates. They employ two basic methods of finding females: patrolling and perching. Patrolling males such as swallowtails fly in areas where females are likely to be present, seemingly searching every square foot of habitat for a resting potential mate. They may find her on a hilltop, a treetop, a forest opening, or patch of nectar flowers in an open meadow. Males of other butterflies such as *Satyrium* hairstreaks choose a territory and perch quietly on tree trunks or low leaves in hopes that a female might enter. They will fly out to investigate virtually any moving object and defend their territory aggressively against other males. When a conspecific male flies into the area, an aerial joust follows. In a dizzying upward spiral flight the two males batter one another. Usually the original male, missing a few scales, returns to reclaim the territory.

Male butterflies identify females by color and scent. When a male encounters a female of the same species, a courtship begins. Courtship behavior is often similar in closely related species. Typically, the male wafts phero-mones in her direction. After a few minutes of courtship, a receptive female will land. The suitor then quickly lands and walks beside her. After bending the tip of his abdomen to meet hers, he firmly grasps her with a pair of valves (claspers). Once joined, the pair faces in oppo-site directions for up to several hours. If frightened, the female may fly off carrying the attached male. Males will mate with several females during their lifespans, females usually once. A female that has already mated will reject a courting male. She typically drops into low vegetation, elevates her abdomen, and rapidly quivers her wings. This rejection dance informs the male she has mated; he usually flies away.

Mating pairs. On the top, a pair of Eastern Tailed-Blues situated in the usual mating position. On the bottom, a pair of American Coppers showing a successful courtship.

Female Summer Azure ovipositing on wingstem floral buds.

Egg Laying and Hostplants

After mating, female butterflies tirelessly search for hostplants on which to lay eggs. Each female must select particular plants that the larvae will eventually eat. Most species have a restricted range of plants that their larvae tolerate. Females recognize these hostplants by their visual appearance and chemical makeup. A great majority of plants have evolved poisonous chemicals that protect them from herbivores. However, butterflies have evolved mechanisms to detoxify poisons in their particular hostplants. After close visual inspection of a potential hostplant, the female tastes it with sensory receptors in her feet. If she recognizes the plant's chemistry as acceptable, she curls the tip of her abdomen downward, touching the plant, and deposits an egg. Each egg is secured firmly to the plant by glue supplied by the female's ovipositor. Most species conceal their eggs on the undersides of fresh, tender leaves. Many swallowtails attach their eggs on the upper sides of fully developed leaves. Many hairstreaks lay eggs on woody twigs bearing next year's leaf buds, and azures lay eggs on unopened floral buds. Mourning Cloaks lay their eggs in ring clusters around twigs of their host.

Butterflies nectaring. On left, Coral Hairstreak and on right, Orange Sulphur.

Feeding

Beyond their primary tasks of mating and laying eggs, adult butterflies must also obtain nourishment and escape from predators. Most butterflies are dependent on floral nectar for nutrients (sugars, amino acids) and hydration. When seeking nectar, the butterfly is guided by the visual and ultraviolet spectra of flowers to inspect the plant closely. Its antennae, palpi, tarsi, and proboscis have sensory receptors that direct it to the nectar. Throughout feeding, butterflies grip flower petals with their legs and extend their proboscises into the bases of the flowers to attain nectar. Most species hold their wings still during feeding, but swallowtails constantly flutter their wings. In return for nectar, butterflies help distribute pollen between plants. In addition to floral energy sources, woodland butterfly species such as anglewings and hackberry butterflies are strongly attracted to sap flowing from wounded trees. Also, virtually all brushfoot butterflies will alight on manure, urine, bird droppings, decaying animals, and fermenting fruit. Males of swallowtails, sulphurs, blues, and skippers frequently visit damp sand or mud to siphon salts from the ground, principally sodium. This activity is known as "puddling" or sometimes "mudpuddling."

Predators

All butterfly life stages are vulnerable to attack by predators. Early stages are constantly threatened by parasitoids, a group of small wasps (especially Braconidae and Ichneumonidae) and flies (Tachinidae). These parasitic insects lay their own eggs on or within butterfly eggs, larvae, and pupae. The larvae of the parasitoids devour the butterfly host. Most of these parasitoids are quite specific in their menu of target species. They seek prey within a narrow range of closely related butterflies.

On the left, a Harvester puddling. Notice the small water droplet adhering to the tip of the abdomen. After fluids are taken in through the proboscis, they are emitted at the tip of the abdomen. On the right, an Arctic Skipper captured by a crab spider whose color blends in perfectly with the flower.

Adult butterflies are attacked by spiders, predatory insects, lizards, and birds. Roaming jumping spiders (Salticidae), ambushing crab spiders (Thomisidae), and web-spinning spiders (Araneidae) capture large numbers of adult butterflies. Dragonflies (order Odonata) take out many adult butterflies around wetlands. Mantids (Mantidae), stink bugs (Pentatomidae), ambush bugs (Phymatidae), and assassin bugs (Reduviidae) also take a toll.

Protective Coloration

The perils of the insect world are many, the biggest of which is being eaten. Adult butterflies have a number of defenses that save them from harm. Most important among them are cryptic coloration, evasive behaviors, and chemicals. Camouflage (or *crypsis*) is found in many butterflies. This type of coloration serves to obscure the butterfly in its natural environment. Good examples are the undersides of tortoiseshells and anglewings, which resemble dead leaves and tree bark (see example below). Disruptive patterns such as stripes also break up the body outline, helping adults to blend with their background. Good examples are the wings of swallowtails, which are boldly marked with rows of discontinuous colors. These blend with shadows, making the butterfly less perceptible.

Using a different strategy, some butterflies display frightening coloration such as false face patterns. Several species have large "eyespots" on their wings (see example below), which startle lizards and birds, giving the butterfly a brief moment to escape. Many lycaenid butterflies such as hairstreaks have small eyespots and slender tails resembling antennae at the tip of their hindwings, creating the impression of a false head. At rest, these butterflies frequently rub their hindwings together, a behavior that deflects a predator's attention away from the true head and toward the false head on the wings. If the predator such as a jumping spider strikes the wrong end, a small piece of wing may be all that the butterfly loses.

Some butterflies are poisonous and deliberately display their toxic properties with bright patterns of red, yellow, orange, and sometimes iridescent blue. This warning, or aposematic, coloration informs predators that they are not edible. Their poisons are most effective on birds. The body of the Pipevine Swallowtail is impregnated with aristolochic acid acquired from pipevine leaves during larval life. This chemical makes birds regurgitate. Young birds that taste a few Pipevine Swallowtails quickly learn to avoid them after becoming ill. Some edible non-toxic butterflies have craftily evolved color patterns very similar to those of distasteful species. For instance, the female Black Swallowtail, the dark female form of the Tiger Swallowtail, and others mimic the distasteful Pipevine Swallowtail. A system in which edible species resemble a distasteful species is referred to as Batesian mimicry. In another twist, the adult Monarch is chemically protected by glycosidic substances (cardenolides) acquired during consumption of milkweeds in larval life. Its lookalike, the Viceroy, is also distasteful due to salicylates acquired from willows in larval life. The distasteful comimics enhance the effect of their bright orange-and-black coloration in a relationship known as Müllerian mimicry.

Polymorphism

Butterflies frequently appear in multiple forms within the same species. The best example is sexual dimorphism ("two forms"), where the male and female of a given species do not look alike. Apart from similar color, the two sexes may differ in size, color patterns, secondary sexual characteristics, and behavior. Sexual dimorphism is widespread among the butterflies and is genetically determined. Females in general are larger, more cryptic, and secretive. These features are designed to increase the female's longevity and her chances of laying eggs before expiring. The longer she lives, the better it is for the species' future success. Sexual dimorphism is present in all families and is quite prominent in certain species of the Pieridae, Lycaenidae, and Nymphalidae.

Environmental conditions may also affect the appearance of butterflies. Although this is a type of polymorphism ("many forms"), it is often referred to as *polyphenism* ("many types"), since it is not an inherited genetic trait. Rather, it is a phenomenon where two or more distinct forms, or phenotypes, are produced by the same genotype, usually from environmental cues. Circumstances such as dry seasons, short day lengths, and cold temperatures may cue hormonal pathways that facilitate this type of transformation. It is quite prominent in certain species of the Pieridae. It is also quite noticeable in the anglewings, where overwintering adults and summer adults are dissimilar (seasonal polymorphism). This nongenetic environmental adaptation optimizes the species' camouflage for two very different seasons.

Butterfly Diversity and Habitat

Species diversity is a measure that incorporates both the number of species and the relative abundance of individuals per species in a given area. Tropical and subtropical areas have the greatest species diversity on Earth. There are several factors that contribute to their rich diversity. For example, the climate enables many more generations of butterflies per year, and the forests are subdivided into many different habitats, each with its own mix of plants and butterflies. A habitat is the natural area where a plant or animal normally lives. Many butterflies are found only in specific places such as a single habitat, while others are distributed broadly across many different habitats.

An awareness of habitats and physiographic regions of the state can guide the butterfly enthusiast in where to look for butterflies. Bedrock, soil conditions, and the vegetation of a particular area have strong influences on its butterfly residents. For instance, shallow soils overlying limestone have high lime content and plants growing there will have a strong affinity for lime. The eastern red cedar, a calciphile tree, readily grows in these soils and the greatest concentration of the Juniper Hairstreak, whose larvae feed only on this tree, follow its distribution. Various types of wetlands have strongly associated butterflies: sphagnum bogs with cranberry provide homes for Bog Coppers; marshes with sedges support sedge skippers and browns; streamside swales with willows often have Viceroys and Acadian Hairstreaks. There are dozens of different kinds of natural communities in Pennsylvania. Making a list of butterflies found in specific habitats is a valuable step in learning about butterflies. Woodlands, mountaintops, old fields, barrens, roadsides, marshes, bogs, and parks are instructive places to explore.

Physiographic Provinces of Pennsylvania

Pennsylvania's landscape has been formed through millions of years of geological processes, including the formation of mountains, sedimentation processes, erosion, and glaciations. The Commonwealth is divided into six physiographic provinces that reflect the variety of the rocks underlying them and affecting the types of soils, plants, and insects that live within them.

Coastal Plain Province: A portion of the Atlantic Coastal Plain is represented by a narrow strip of land in the southeast corner of the state. The topography is low and relatively flat. Both fresh and saltwater marshes occur in the region. The area at one time supported a healthy population of the Salt Marsh Skipper and also received strays from the outer Coastal Plain, such as Arogos and Dotted Skippers. The area today receives southern migrants following the coast like the Queen, Gulf Fritillary, and Brazilian Skipper.

Piedmont Province: Inland from the Coastal Plain, the Piedmont is characterized by gently rolling hills and broad valleys. Fertile soils support a mixed broadleaf forest of oak, hickory, beech, maple, tulip tree, cherry, and flowering dogwood. Butterflies are relatively numerous in the woodlands and adjacent meadows. Many wide-ranging butterflies inhabit the area, such as Tiger and Spicebush Swallowtails, Clouded and Orange Sulphurs, Spring Azure, Eastern Tailed-Blue, Banded Hairstreak, Great Spangled Fritillary, Little Wood Satyr, and skippers. Local igneous intrusions have left serpentine outcrops known as barrens. The herbaceous flora of these barrens is unique, featuring tall grasses with scattered junipers and scrub oaks. Typical butterflies occurring there are the Falcate Orangetip, Juniper Hairstreak, and Dusted Skipper.

Blue Ridge Province: This narrow strip of a well-forested, flat-topped mountain ridge terminates in south-central Pennsylvania. Locally it is called South Mountain. The natural vegetation is a dry Appalachian oak forest with an understory of mountain laurel and lowbush blueberry. Typical butterflies residing there are Appalachian Tiger Swallowtail, Brown Elfin, Northern Spring Azure, and duskywing species.

Ridge and Valley Province: Through the heart of the state occurs a belt of southwest-to-northeast-trending ridges and intermontane valleys, known locally as the Appalachian Mountains. The ridges support a dry Appalachian oak forest, not unlike the Blue Ridge Province, but also feature shale barrens with unique flora. The valley flora resembles that of the Piedmont. Unique butterflies of this province include Appalachian Tiger Swallowtail, Falcate Orangetip, Olympia Marble, Northern Metalmark, Henry's Elfin, Northern Spring Azure, Silvery Blue, and Appalachian Grizzled Skipper.

Appalachian Plateau Province: The northern and western boundary of the Ridge and Valley Province is demarcated by a sharp escarpment known as the Allegheny Front. It marks the beginning of a vast elevated plateau that slopes gently to the west. Over eons the plateau has been thoroughly dissected into rugged topography by river systems. The northern high-elevation section of the province features a northern hardwood forest of spruce, pine, hemlock, beech, maple, oak, and cherry mixed with wetlands. Typical butterflies of this section are the Canadian Tiger Swallowtail, West Virginia White, Bog Copper, Acadian Hairstreak, Northern Spring Azure, Cherry Gall Azure, Atlantis Fritillary, Harris' Checkerspot, Northern Crescent, Green Comma, White Admiral, Eyed Brown, Arctic Skipper, Dion Skipper, and Two-spotted Skipper. The southern "tongue" of this section contains diffuse northern floral elements and very few of the northern butterflies; rather, the butterfly fauna of the tongue looks more like that of the Ridge and Valley Province. The western low-elevation section of the province has been heavily dissected; it appears to be a series of high hills and steep valleys near major rivers and low rolling hills elsewhere. The area features a mixed deciduous forest dominated by oaks and hickories; flowering dogwood occurs in the understory with a herbaceous layer of spring flowers including broadleaf toothwort; Virginia pine, pitch pine, and shortleaf pine grow on ridges. Many wide-ranging resident and migrant butterflies, common to southern regions of the state, occur in the area. Typically, good numbers of the West Virginia White appear in the spring. Unique butterflies like the Dusky Azure, Swamp Metalmark, and Diana Fritillary once occurred in this section. These species likely arrived in southwestern Pennsylvania via extensions of populations in the Ohio River valley.

Central Lowlands Province: This thin band of the eastern Great Lakes lowlands consists of lacustrine deposits and old beach outlines on a nearly level terrain adjacent to Lake Erie. The natural vegetation is beech-maple forest, except for an eastward trending peninsula (Presque Isle) created by sand deposits. The lake acts as a barrier for southern migrants, which often turn eastward and continue their migrations through the region, such as Giant Swallowtail, Checkered White, Little Yellow, Dainty Sulphur, Marine Blue, Hayhurst's Scallopwing, and Ocola Skipper.

Collecting

Part of our obligation and duty as naturalists is to preserve our natural heritage. While doing so, we must constantly increase our knowledge of butterflies. Developing a reference collection is one of several legitimate activities enabling naturalists to document regional diversity and variability of butterfly species. Great museums all over the world house millions of butterflies and assist in deciphering their vast array of anatomical features, adaptations, and genetic changes over time. In our own lifetime a personal collection can serve as the basis for a local checklist and a means of introducing children and adults to their natural environment. Serious avocational collectors may interact with professional lepidopterists to augment institutional collections and to help resolve taxonomic questions. Collecting butterflies requires discipline in the field and at home. Natural butterfly populations should be judiciously sampled and never depleted.

Most butterfly nets weigh less than one pound and are composed of a fine-mesh polyester

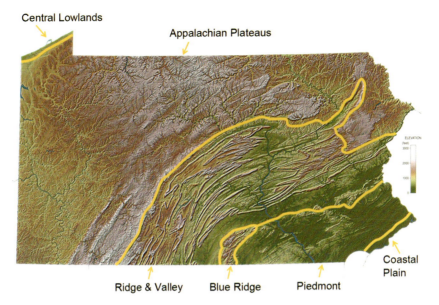

Central Lowlands

Appalachian Plateaus

ELEVATION
(feet)
3000
2000
1000
0

Coastal Plain

Ridge & Valley   Blue Ridge   Piedmont

Shaded relief map of Pennsylvania showing physiographic province boundaries in yellow.

bag supported by a noncorrodible aluminum ring (twelve to fifteen inches in diameter) and a handle (three to five feet in length). Once netted, if a butterfly is being retained as a collection voucher, it is recommended that it be stunned by firmly pinching the thorax (wings folded up) using your thumb and opposing index finger. The butterfly may then be placed in a clean glassine envelope and transported home or to the lab. Ultimately, to be of scientific value, the butterfly must be pinned and spread with wings extended. For these purposes, biological supply houses offer pinning boards, entomological pins, labels, and storage drawers and cabinets. All known data should be recorded with the specimens (species name, location, date, collector, habitat, hostplant, observations of behavior, and biological interactions). The data should be made available to appropriate interested parties upon request. The collection should be protected from physical damage and offered to an appropriate scientific institution when it can no longer be properly managed.

Butterfly Watching

In recent decades butterfly watching has become a major recreational activity. Observing with the unaided eye takes patience and years of experience. It can also be frustrating when trying to identify small dull species like grass skippers. Technological advances have assisted in the pursuit. Close-focusing binoculars and a respectable camera are essential components of the butterfly watcher's tool kit. These devices bring the butterfly up close for inspection without alarming it. The ideal binoculars are those that magnify six to seven times (6×–7×) life size. The size and weight of the binoculars should feel comfortable in your hands and should not cause fatigue after many hours of use. Acquiring a collection of butterfly images is a far less complicated and less expensive task than it was two decades ago. Modern digital cameras are both light and adaptable, permitting photographers to acquire excellent photographs of butterflies. It is essential to get the right equipment. A single-lens reflex (SLR) camera and a macro lens of 100 mm to 200 mm range make it easier to get sharp, eye-catching photos without being too close to the subject. (Even though they are not recommended, small point-and-shoot digital cameras come in handy in a pinch when you're without your SLR camera. Acceptable photos can be obtained, but you must be within inches of the subject.) A macro lens in the 50 mm range is excellent for photographing immature stages (eggs, larvae, pupae). Lastly, a net and jar are very useful for the butterfly watcher. Butterflies can be netted and placed in a clear bottle for extended periods of observation. This practice helps confirm difficult identifications and is used by educators when introducing beginners to butterfly study.

Butterfly Gardening

Gardening for butterflies is an excellent method for butterfly watching. It is possible to attract them to your own yard, whether you live in a big city, a suburb, or the country. When planning a butterfly garden, we recommend selecting flowering plants that supply nectar through all flight seasons from spring to fall. Opt for flowers that are fragrant and high in nectar content. Keep in mind that flowers that have short nectar tubes permit access to the greatest range of butterfly species. The following flowers are examples that can be easily grown in Pennsylvania to attract butterflies: asters, black-eyed Susan, butterfly bush, butterfly weed, clovers, coneflowers, daisies, dame's rocket, honeysuckles, joe-pye weed, milkweeds, mints, phlox, sedums, thistles, and yarrows. Plants such as roses, lilies, and geraniums have little nectar and are not useful in a butterfly garden. When planning your garden, make sure it will get as much sun throughout the day as possible. Nearby shelters, such as bushes and small trees, are also recommended to provide protection from wind and rain during adverse weather. One of the most satisfying ways to learn about butterflies close to home is to follow their life cycle through egg, larva, and pupa stages. Female butterflies sometimes wander long distances looking for larval hostplants. Planting a few of these will increase the chances that they will be attracted to your garden. Mustards and parsley will bring in species of whites and Black swallowtails to lay eggs. Milkweeds will attract Monarchs; beans will attract sulphurs; and thistles or nettles will draw in some brushfoot butterflies. If there is concern about damage to your home-grown vegetable crop, you may avoid planting these types of plants. Pay special attention to blues and hairstreaks that regularly visit garden flowers for nectar; they may be laying eggs on neighborhood oak, hickory, and cherry trees, or even the clovers in your yard. For added attraction, consider discarding your rotting fruit in the garden. Anglewings, Mourning Cloaks, emperors, admirals, and snouts are drawn to this resource like magnets. They may drink from it for hours.

Changing Butterfly Distribution and Conservation

Butterfly populations are in a constant state of flux. Changes in their ranges and sizes reflect adjustments to changing climates and the composition of natural communities. A natural ebb and flow, including extinction, is part of life on Earth. This is certainly true over long periods of time considering the effects of glacial epochs, solar cycles, and prominent geologic events. All of these events have had profound effects on butterfly evolution. Over shorter periods of time, such as a human lifetime, we have begun to pay attention to anthropogenic (human-created) disturbances and their effect on shaping the natural world around us. In most cases, human disturbances have been destructive to natural habitats and the species within them. Before European settlers arrived, Pennsylvania was mostly a mosaic of forests and scattered savannas (open prairielike areas). Logging, mining, modern agriculture, residential development, dams, canals, roads, railroads, and utility lines have altered the topography and vegetation of significant parts of the state. Fortunately, nearly 60% of Pennsylvania remains forested. Most Pennsylvania butterflies occupy wide ranges throughout the continent and are in no present danger. To date, no butterfly species found in Pennsylvania has become extinct. However, in the last three decades, a broad public concern has emerged for the conservation of butterflies and their local habitats. By our estimation, 14% of our resident butterflies either no longer occur in the state or are critically imperiled.

There are several explanations for why butterflies should be conserved, but foremost among these is that they are quality indicators of a healthy ecosystem. Where butterfly colonies remain stable in their natural habitat, they serve as signs of a healthy environment. Some naturalists believe that laws preventing butterfly collecting are needed to protect vulnerable species. A blanket ban on collecting all butterfly species has never been demonstrated to aid the survival of a single population. Rather, the greatest danger to imperiled butterfly populations is the steady destruction of their habitats. The all-encompassing best method to protect vulnerable species is the preservation of their habitat. Restoring habitats, though useful, cannot always recreate what has been lost. Threatened habitats where the greatest numbers of imperiled butterfly species are found (wetlands, savanna) are in particular need of protection. Many of these habitats are currently protected, since they are in state parks, gamelands, forests, and national forests. Yet year-to-year surveys of butterflies and annual review of management practices are important to achieving optimal survivals. Despite pressures from human populations, butterfly habitats can be preserved; a prime example is the last remaining habitat for the Regal Fritillary in the eastern half of the continent at Fort Indiantown Gap, Pennsylvania.

Every effort, no matter the magnitude, helps to ensure that butterflies can still be enjoyed. Besides backyard butterfly gardening, one can volunteer to assist local land managers with time-consuming tasks, such as planting nectar flowers and larval hostplants, removing invasive plants, advising when and how to mow, and assisting with controlled burning where needed.

A standardized numerical scale is often used by a number of organizations as a means of denoting the conservation status of a species. In the species account section of this guide, an S-ranking is provided for each species. The definitions of these numerical ranks are as follows:

S1 Critically imperiled. Critically imperiled because of extreme rarity (often ≤ 5 occurrences) or because of some factors such as very steep declines making it especially vulnerable to extirpation from the state.

S2 Imperiled. Imperiled because of rarity due to a very restricted range, very few populations (often ≤ 20), steep declines, or other factors making it very vulnerable to extirpation from the state.

S3 Vulnerable. Vulnerable due to a restricted range, relatively few populations (often ≤ 80 occurrences), recent and widespread declines, or other factors making it vulnerable to extirpation.

S4 Apparently secure. Uncommon but not rare; some cause for long-term concern due to declines or other factors.

S5 Secure. Common, widespread, and abundant.

S#S# Range rank. A numerical range (e.g., S3S4) is used to indicate any range of uncertainty about the status of species or community.

SNA Not applicable. Rank is not applicable because the species is not a suitable target for conservation activities (e.g., migrant, stray, vagrant).

SH Historical (possibly extirpated). Species or community occurred historically in state, and there is some possibility that it may be rediscovered. It may not have been verified in past twenty to forty years. This rank is reserved for species or communities for which some effort at relocation has occurred.

SX Presumed extirpated. Species or community is believed to be extirpated from the state. Not located despite intensive searches of historical sites and other appropriate habitat, and virtually no likelihood it will be rediscovered.

B Breeding. Conservation status refers to breeding in state. This rank is typically used with annually occurring migrants that breed in state, particularly at staging areas or concentration spots where the species might warrant conservation attention.

* Tracked. Species tracked/monitored by Pennsylvania Natural Heritage Program (PNHP).

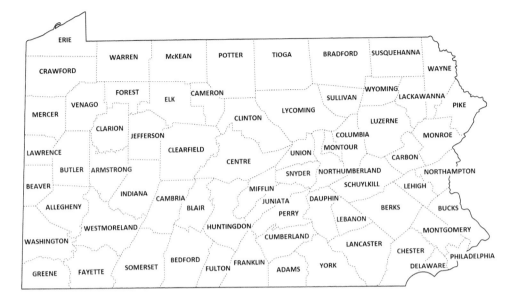

One hundred and fifty-six (156) species have been recorded in Pennsylvania. Each of these is covered in the following species accounts. Scientific names have been provided to the appropriate subspecies level, though subspecies are rarely discussed in the text. The only exceptions are when two subspecies are both found in the state. In addition to species recorded in Pennsylvania, another ten species are listed that for a variety of reasons might occur in the state and should be looked for.

Information for each recorded species begins with photos of dorsal and ventral sides of both genders, as well as seasonal forms and major varieties where important. The photos are supplemented with text material denoting their distinguishing marks and typical behavior. Distribution maps and flight phenograms provide evidence of where and when they fly. Finally, details of larval hostplants and preferred habitat give further clues to identifying them and appreciating their life history.

All species presentations follow a specific format. An annotated version of a typical account is found on the next two pages.

average wing span

Distinguishing marks: (1) Coppery orange field in dorsal and ventral forewing with black spots; (2) prominent submarginal orange band on dorsal hindwing; and (3) thin submarginal red-orange line on gray ventral hindwing.

Typical behavior: Flies close to ground. Often found nectaring on small wildflowers. Males territorial, attacking intruders entering their territory.

Habitat: Open areas such as meadows, fields, roadsides, and utility line cuts.

Larval hosts: Sheep sorrel (*Rumex acetosella*).

Abundance: Common, with localized colonies; seemingly absent in some years, common in others. S5

Remarks: Three broods, occasional fourth. Overwinters as partially grown larva. Similar to female Bronze Copper (p. 72), but significantly smaller. Orange submarginal line on ventral hindwing of Bronze Copper is thicker (bandlike). Pockets of population in northeastern portion of state with forewing black spots enlarged and fused (form "*fasciata*").

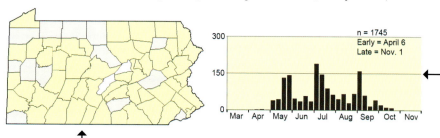

The common name of the species, with the common name generally from the North American Butterfly Association in their Checklist of North American Butterflies Occurring North of Mexico.

The scientific name of the species to the subspecies level with the scientific names from *A Catalogue of the Butterflies of the United States and Canada* by Johnathan P. Pelham being followed.

The author of the species and the date it was authored. The author of a species is the individual given credit for its initial description. As with the scientific names *A Catalogue of the Butterflies of the United States and Canada* by Johnathan P. Pelham is followed. A parenthesis around the name and date indicates in the original description a different genus name was used.

Photographs of both the dorsal and ventral side of a typical male, ♂, and female, ♀, are presented.
Arrows pointing to specific markings are overlaid on the photographs. These arrows numbered corresponding to the numbers given in the section Distinguishing marks. Not all numbered items in Distinguishing marks will have a corresponding arrow. Often characteristics of a broad area have no corresponding arrow.

Two-headed arrow denotes the average wing span.

Specific marks allowing for identification of the species.

A short description of the behavior of the species. Examples being nectaring, attracted to dung, carrion, etc., strong flier, weak flier, territorial, etc. Anything that might help one identify the species based on something other than the species' markings.

While few butterflies can be completely tied down as to one specific habitat there is a large difference between the open areas of the pierids and the forest occupancy of the Northern Pearly-eye.

Only larval hostplants recorded from Pennsylvania are listed, unless noted otherwise.

Besides a general description a numerical value, the official state ranking, is given. Definitions of these numerical ranks is given in the Introduction section.

Miscellaneous remarks covering similar species, historical sightings, or any other comment that would help in the identification or understanding of the species.

Over 40 years of data is presented showing a weekly break-down of sightings, the number of sightings, the first and last sighting of the year.

A Pennsylvania map with the county boundaries shown is given for each species. If a species has been recorded within a county in the last twenty years this county is colored yellow, if recorded from the county but not within the last twenty years then colored gray, otherwise uncolored.

A representative of each of the six families of butterflies and skippers found in Pennsylvania are shown here. On the previous page, clockwise from the top are: Papilionidae—Pipevine Swallowtail, Lycaenidae—Bronze Copper, Hesperiidae—Arctic Skipper, and Pieridae—Orange Sulphur. On this page, from left to right are: Riodinidae—Northern Metalmark and Nymphalidae—American Snout.

Species Accounts

Three subfamilies make up the family Papilionidae. Only one of these subfamilies has representatives found in Pennsylvania. This is the subfamily Papilioninae, commonly known as the true swallowtails. Nine species have been recorded from Pennsylvania. These nine are further separated into tribes, with one member of the Aristolochia swallowtails (Pipevine Swallowtail), one member of the kite swallowtails (Zebra Swallowtail), and the remainder in the fluted swallowtails. Most are year-round residents and among our most common butterflies, while one is a rare migratory stray (Palamedes Swallowtail).

Members of the family Papilionidae are the largest butterflies in the state. They are strong fliers, colorful, and avid nectarers. They are easy to recognize by their signature feature, long tails extending from the hindwings. This character, which resembles the long tail of a swallow, is the reason for the common name. These tails divert bird attacks away from critical body parts such as the butterfly's head and body, allowing the butterfly to escape.

Wing and body colors of the Papilioninae vary from black to white to yellow with black stripes. One black-colored swallowtail with a blue iridescent sheen (Pipevine Swallowtail) is distasteful and avoided by birds. This foul-tasting swallowtail is mimicked by several other swallowtails (Spicebush Swallowtail, female Black Swallowtail, dark females of Eastern Tiger and Appalachian Tiger Swallowtails) and at least one non-swallowtail (Red-spotted Purple). Through this similarity, these mimics garner protection from bird attacks, especially in areas where the Pipevine Swallowtail is common.

Ventral side of Pipevine Swallowtail

Our swallowtail species fly in multiple broods from spring to fall. The best opportunities to see them are in spring and summer. The hostplants that these swallowtails use fall along generic lines. The Pipevine Swallowtail and Zebra Swallowtail have single hosts, pipevine and pawpaw, respectively. The remaining fluted swallowtails use leaves of various forest trees and shrubs, plus a few wildflowers. Their eggs are generally spherical and undecorated. Larvae look very much like bird droppings in early stages and small snakes with fake eyespots on the thorax in later stages. The chrysalises are brown- and green-colored, resembling pieces of bark or leaves. They hang upright supported by a silk girdle.

In spring and summer, after rain showers, look for males sipping moisture from wet soil. They often congregate in large groups known as "puddle parties."

Dorsal side of male Eastern Tiger Swallowtail

A puddle party with four species of swallowtails (Palamedes, Zebra, Eastern Tiger, Black)

average wing span

Distinguishing marks: Dorsal forewing with (1) two rows of cream-colored spots; female often with only partial inner row. Ventral forewing with (2) two rows of yellow-orange spots. Ventral hindwing with (3) an unbroken median row of orange spots on ventral hindwing with an extra orange-yellow spot in discal cell. Female (4) with blue on dorsal hindwing, mimicking the Pipevine Swallowtail.

Typical behavior: Females are most often found nectaring and males are often found puddling.

Habitat: This species can be found in almost any type of open area. It is often found in disturbed areas such as gardens, lawns, weedy fields, and parks, especially gardens with parsley, carrot, or dill species.

Larval hosts: The larvae feed on species of Apiaceae, commonly known as the parsley and carrot family. They use domesticated species such as parsley, carrot, dill, and garden rue, as well as several wild species of plants in this family.

Abundance: A common resident throughout the state. S5

♀

Remarks: Two broods with occasional partial third brood. Overwinters as pupa. The female Black Swallowtail with prominent blue scaling on its dorsal hindwing mimics the the poisonous Pipevine Swallowtail (p. 30). The Giant Swallowtail (p. 36) somewhat resembles the Black Swallowtail; however, its larger size and yellow venter easily distinguish it from the Black Swallowtail.

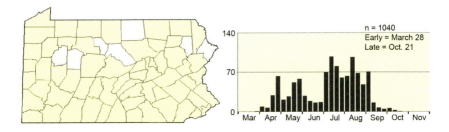

n = 1040
Early = March 28
Late = Oct. 21

average wing span

Distinguishing marks: Very large. Dorsal side with (1) two rows of yellow spots, (2) merging toward apices of forewings. Yellow venter interrupted by blue median line on hindwing.

Typical behavior: Females are most often found nectaring and males are often found puddling.

Habitat: Prefers open areas such as fields, clearings in woodlands, gardens, etc.

Larval hosts: In southern states, larvae feed on citrus trees, making it an agricultural pest. In Pennsylvania, the main hostplants are the hoptree (*Ptelea trifoliata*) and common prickly ash (*Zanthoxylum americanum*), as well as members of the citrus or rue family (Rutaceae).

Abundance: A migrant with occasional temporary colonization. Generally uncommon, but may appear in significant numbers in late summer as migrants. S4B

♀

Remarks: Two broods. Overwinters as pupa. Its large size is enough to distinguish this species from any other Pennsylvania swallowtail. While the butterfly is in flight the ventral side makes a very strong impression. The weak numbers of the early brood indicate that there is a small overwintering population. The larger second brood is considered to be the result of southern individuals migrating north in summer. The full story of the natural history of this species in Pennsylvania is unclear.

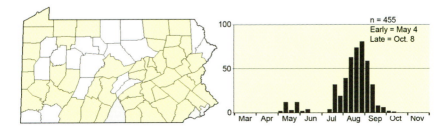

n = 455
Early = May 4
Late = Oct. 8

average wing span

Distinguishing marks: (1) Narrow black band along inner margin of hindwing. (2) Submarginal yellow band on ventral forewing separated by black veining. (3) Junction of yellow and black fields on ventral hindwing is scalloped and (4) yellow-orange spots along outer edge of ventral hindwing are crescent-shaped. Females have (5) blue on the dorsal hindwing and (6) first submarginal spot on ventral hindwing that is bigger and generally shaped like the letter *D*.

Typical behavior: Strong fliers. Found soaring high in trees, fluttering and nectaring, or puddling in large numbers on wet ground (males).

Habitat: Open areas such as fields, roadsides, gardens, along riverbanks, etc., but not far from deciduous forests.

Larval hosts: Wide variety of tree leaves, including black cherry (*Prunus serotina*), tulip tree (*Liriodendron tulipifera*), hoptree (*Ptelea trifoliata*), and white ash (*Fraxinus americana*).

♀

5

6

Abundance: Abundant throughout the state. S5

Remarks: Two broods. Overwinters as pupa. Hybridizes with Canadian Tiger Swallowtail (p. 40) in a broad zone in high plateaus of northern Pennsylvania. There is a dark form of the female of this species. This form is more common in the second brood and is part of the Pipevine Swallowtail (p. 30) mimicry ring.

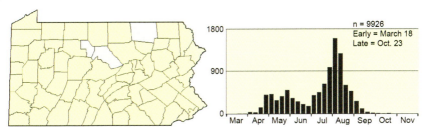

n = 9926
Early = March 18
Late = Oct. 23

1800

900

0

Mar Apr May Jun Jul Aug Sep Oct Nov

average wing span

Distinguishing marks: Similar to but smaller than Eastern Tiger Swallowtail (p. 38).
(1) Wide black band on inner margin of hindwing. (2) Submarginal yellow band on ventral
forewing not separated by black veining. (3) Junction of yellow and black fields on ventral
hindwing is nearly straight, not scalloped, and (4) yellow-orange spots along outer edge of
ventral hindwing are more rectangular than crescent-shaped. Females have (5) blue on the
dorsal hindwing, but less than female Eastern Tiger Swallowtail, and (6) first submarginal
spot on ventral hindwing that is bigger than the same spot on the male and generally shaped
like the letter *D*.

Typical behavior: Strong flier. Both sexes often found nectaring. Males often found puddling.

Habitat: Openings and edges in northern forests.

Larval hosts: Leaves of aspens (*Populus*) and black cherry (*Prunus serotina*).

Abundance: Abundant, but limited to the northern plateaus of the state. S5

Remarks: One brood. Overwinters as pupa. The Canadian Tiger Swallowtail in Pennsylvania is a hybrid form (*canadensis* × *glaucus*), yet its overall appearance and larval hosts are typical of the northern species. Its single brood is broken into an "early" flight and a "late" flight, as seen in the flight phenogram. Further research is needed to determine if these flights are diverging genetically. A black female form generally does not occur.

average wing span

Distinguishing marks: Largest of the tiger swallowtail species and wings more angular than the others. (1) Intermediate-width black band along inner margin of hindwing. (2) Submarginal yellow band on ventral forewing intermediate between that of Canadian and Eastern Tigers (pp. 40, 38). (3) Junction of yellow and black fields straight and least scalloped of all three tiger species. (4) Yellow-orange spots along outer edge of ventral hindwing more rectangular than crescent-shaped. Females with (5) less blue on the dorsal hindwing than female Eastern Tiger; and (6) first submarginal spot on the ventral hindwing that is bigger than the same spot on the male and generally shaped like the letter *D*. At puddle parties of Eastern and Appalachian Tigers, the latter noticeably stands out due to its larger size.

Typical behavior: Strong flier. Both sexes nectar avidly. Males often found puddling.

Habitat: Openings and edges of deciduous forests of southern Ridge and Valley Province.

♀

Larval hosts: Black cherry (*Prunus serotina*) is the only currently known host.

Abundance: Locally common. Easily over-looked. S4

Remarks: The Appalachian Tiger Swallowtail is univoltine and flies between the first and second brood of the Eastern Tiger Swallowtail. Overwinters as pupa. It occurs in the southern region of the state, while the Canadian Tiger is found in the northern region. These two species meet in only one county (Centre County), but even there they do not overlap. It is hypothesized that this species originated at high elevations of southern Appalachian Mountains in previous glacial periods.

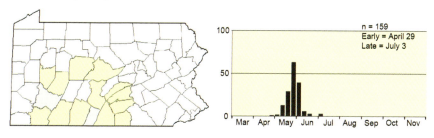

n = 159
Early = April 29
Late = July 3

In general, size differences can be a first step in distinguishing the three species of tiger swallowtails, with smallest to largest in this order: Canadian, Eastern, Appalachian. Three additional characteristics are covered below using magnified side-by-side views. The order from left to right in each case is Eastern, Canadian, and Appalachian.

The ventral forewings of the tiger swallowtails feature a yellow submarginal band. This band is broken into separate spots by black veining in the Eastern Tiger, but unbroken by black veining in the Canadian. The Appalachian banding is intermediate. Inside the submarginal band is a black band with a small amount of pale overscaling. In the Eastern Tiger this overscaling is faint. In the Canadian and Appalachian Tigers there is much more overscaling, making this band appear as a silver-gray band to the naked eye.

The black band along the inner margin of the hindwing is widest in the Canadian and narrowest in the Eastern. The width should be judged relative to where the nearest wing vein splits. This location is indicated by the forked black line in the photos.

The junction of the yellow and black field on the ventral hindwing is (1) most scalloped in the Eastern Tiger Swallowtail. This edge is (2) nearly straight on the Canadian and Appalachian. Similarly, the orange-centered, cream-colored marginal spots are (3) more curved in the Eastern Tiger Swallowtail, while (4) more angular, almost rectangular, in the Canadian and Appalachian.

Female Eastern and Appalachian Tiger Swallowtails also have a dark form that mimics the distasteful Pipevine Swallowtail. The proportion of dark form females increases in proportion to the abundance of Pipevine Swallowtails flying sympatrically with the tigers.

Shown to the right are the dorsal and ventral sides of the dark form of the female Eastern Tiger Swallowtail. Some of the distinguishing marks of the yellow form are still visible: (1) A single submarginal row of pale yellow spots on the dorsal forewing and (2) a prominent black stripe seen in the middle of ventral hindwing.

♀

1 →

← 2

An array of partially pigmented dark forms occur in nature. To the right is an intermediate speckled dark form of the female Eastern Tiger Swallowtail.

average wing span

Distinguishing marks: Dorsal hindwing with (1) blue-green iridescence in male and (2) blue in female. (3) Row of pale green spots in submarginal area of dorsal wings. Ventral hindwing with (4) inner row of orange spots interrupted by a blue "rocket."

Typical behavior: A strong flier. Both sexes take nectar. Males puddle.

Habitat: Found in open areas, often in disturbed areas like yards and gardens, but also along edges of woods containing either sassafras or spicebush trees.

Larval hosts: Sassafras (*Sassafras albidum*) and spicebush (*Lindera benzoin*) trees.

Abundance: Abundant throughout Pennsylvania. S5

Remarks: Two broods. Overwinters as pupa. A member of the Pipevine Swallowtail (p. 30) mimicry ring. Look for larvae in silk-lined leaf shelters, which look like small green snakes with large eyes, on host trees.

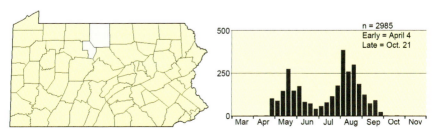

n = 2985
Early = April 4
Late = Oct. 21

average wing span

Distinguishing marks: Dorsal wings with (1) two rows of yellowish spots, becoming a broad band on hindwing; and (2) pale yellow bar at end of discal cell on forewing. Ventral hindwing with (3) thin yellow-orange line near base.

Typical behavior: Strong flier. Avidly nectars. Like other swallowtails, males of this species often found puddling.

Habitat: Generally found in southern swamplands.

Larval hosts: Red bay (*Persea borbonia*) in southeastern states.

Abundance: Very rare stray. SNA

Remarks: Does not breed in state. A common swallowtail in the Southeast. Possibly will be seen more often in Pennsylvania in the future due to warming climate.

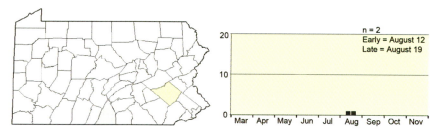

n = 2
Early = August 12
Late = August 19

Four subfamilies make up the family Pieridae. Two of these subfamilies have representatives found in Pennsylvania. They are the subfamily Pierinae, commonly known as the whites, and the subfamily Coliadinae, commonly known as the sulphurs. Three species (Cabbage White, Clouded Sulphur, Orange Sulphur) are among the most common butterflies seen in Pennsylvania. One species (Olympia Marble) is one of the rarest in the state and hasn't been seen in thirty years.

Wing and body colors of the Pieridae vary from white to yellow to orange. These colors are created by pterin pigments believed to be derived from waste products of larval life. In many instances these pigments interact with ultraviolet light, which the human eye cannot see but which butterfly eyes can. The resulting patterns are critical visual cues for mate location and propagation of individual species.

Members of this family range in size from the very small (Dainty Sulphur) to large (Cloudless Sulphur). Some are weak fliers rarely found more than a few feet off the ground and seldom far from their larval hostplants (West Virginia White). Others are strong fliers and migrate annually from southern states to recolonize the state (Sleepy Orange, Little Yellow).

Some members of the Pieridae fly only once a year for a short period of time (univoltine), while others fly nearly continuously via multiple broods (multivoltine). Univoltine species, like the Falcate Orangetip, generally make their brief appearance in spring and then are not seen until the following spring. Multivoltine species, like the Cabbage White and Orange Sulphur, are among the first species flying in the spring and among the last flying in the autumn as winter approaches. These species have so many overlapping broods that it is virtually impossible to determine when one brood ends and another begins. Species that fly through several seasons are subject to a great deal of variation in size and markings due to environmental factors such as temperature and day length.

Above, a Southern Dogface (left) and a Little Yellow (right) are shown in a characteristic pose while nectaring with their wings closed.

The whites primarily use mustards as larval hostplants, while the sulphurs generally use legumes. Eggs are spindle-shaped with thin vertical ribs. Larvae are cylindrical, slender, and green, with some having thin colorful stripes and others being solid green. Larvae are not particularly hairy. However, their thin hairs are hollow and contain deterrent chemicals.

Adults of both sexes are fond of nectar-rich flowers. Males frequently sip moisture from wet soil and can be found in large groups, known as "puddle parties."

Above are two more examples of pierids in their characteristic pose while nectaring with their wings closed: Cabbage White (top) and Orange Sulphur (bottom).

Cabbage White   *Pieris rapae rapae*   (Linnaeus, 1758)

average wing span

Distinguishing marks: Dorsal forewing with (1) black apex; and (2) a single black post-median spot (male); or (3) two black spots (female) in this region. Both sexes with two black postmedian spots on ventral forewing.

Typical behavior: Commonly seen nectaring at flowers in cultivated areas and gardens. Males puddle.

Habitat: Any open area, expanding into forest clearings.

Larval hosts: Many species in mustard family (Brassicaceae), including wild species and cultivated cabbage, cauliflower, broccoli, Brussels sprouts, radishes, etc.

Abundance: Abundant throughout the state. S5

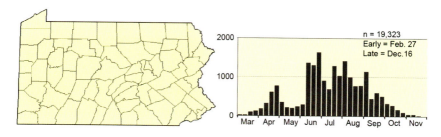

spring / fall form

Remarks: Five or more broods. Overwinters as pupa. Early spring and fall individuals of this species are reduced in size and their markings are much reduced, as shown above. The ventral hindwing is speckled gray-green. If sighted in early spring, one must be careful not to mistake this species for a West Virginia White.

Also known as the European Cabbage Butterfly. This species was accidentally introduced into North America at Quebec in 1860. Thereafter it quickly spread across southern Canada and the United States. By 1870 it was present in Pennsylvania, where it was already reported as an agricultural pest. Farmers and gardeners know the larva as the cabbage worm.

Today this species is likely the most common butterfly in the United States. During Fourth of July butterfly counts of the North American Butterfly Association, it is frequently the most recorded species. Also note the extreme number of sightings recorded in the flight phenogram on the previous page.

Special Topic   Discal Cell

Wing veins are important features that aid in the classification of higher taxa (e.g., families, genera) and in some instances in identifying species. The spaces between veins are called cells. There is one large cell beginning at the base of the wings, which is known as the discal cell. Shown on the right are the discal cells of the Eastern Tiger Swallowtail (p. 38), which have been highlighted to display their location. The "distinguishing marks" section in the species accounts often refers to the discal cell. For instance, note the black bar at the end of the discal cell in the forewing of the Checkered White and the Olympia Marble (pp. 52, 56).

♂   ♀

average wing span

Distinguishing marks: Dorsal forewing with (1) black bar straddling the end of discal cell. Ventral hindwing with green banding pattern, often with faint rosy blush on costal margin.

Typical behavior: Nectars and flies low to the ground.

Habitat: Dry rocky slopes, ridgetops, pipelines, and power line cuts across ridges.

Larval hosts: Rock cresses (*Arabis*).

Abundance: Possibly extirpated. SH*

Remarks: One brood. Overwinters as pupa. The Appalachian population is quite fragmented. The butterfly has not been seen in Pennsylvania in over thirty years. However, it persists in northeastern West Virginia and adjacent areas of Maryland close to the Pennsylvania border.

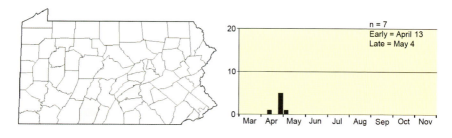

n = 7
Early = April 13
Late = May 4

♂   ♀

average wing span

Distinguishing marks: Forewing apex hooked (falcate). Dorsal forewing apex of male (orange) and female (indistinct). Ventral hindwing with dark green marbling.

Typical behavior: Low, erratic flight in woods.

Habitat: Deciduous woods, dry ridgetops, shale and serpentine barrens.

Larval hosts: Rock cress (*Arabis*) and bittercress (*Cardamine*).

Abundance: Uncommon. Common in southern Ridge and Valley Province. S3*

Remarks: One brood. Overwinters as pupa. The Falcate Orangetip is a harbinger of spring wherever it is found. Just when leaf buds on trees begin to open and the forest takes on a green tinge, this butterfly begins to appear. The males, with their orange-tipped forewings, are a special delight at this time.

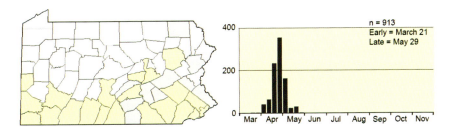

n = 913
Early = March 21
Late = May 29

♂   ♀

average wing span

Distinguishing marks: Light yellow. Dorsum with black borders; (1) female forewing border containing yellow spots. Ventral hindwing with (2) two unequal circles at end of discal cell.

Typical behavior: Frequently nectaring and an active flier, especially on sunny days.

Habitat: Any open area, including fields, pastures, meadows, and gardens.

Larval hosts: Legumes, especially clovers like white clover (*Trifolium repens*) and red clover (*T. pratense*).

Abundance: Common, widespread, familiar butterfly. S5

Remarks: Four to five broods. Overwinters as partially grown larva. Can be confused with the Orange Sulphur (p. 59) and the Pink-edged Sulphur (p. 61). White female form occurs; see Special Topic: "*Alba*" form (p. 60).

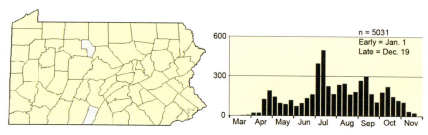

n = 5031
Early = Jan. 1
Late = Dec. 19

♂   ♀

average wing span

Distinguishing marks: Yellow-orange. Dorsal side with black borders; (1) female forewing border containing yellow spots. Ventral hindwing with (2) two unequal circles at end of discal cell.

Typical behavior: Frequently nectaring and active flier, even during cool autumn days.

Habitat: Any open area, including alfalfa fields, pastures, and gardens.

Larval hosts: Legumes, especially clovers, alfalfa, and crown vetch.

Abundance: Abundant, widespread, familiar butterfly. S5

Remarks: Four to five broods. Nondiapausing larvae continue to feed during mild winters. Can be confused with the Clouded Sulphur (p. 58). Orange Sulphur is quite variable; some individuals tend toward yellow ground color. White female form also occurs; see Special Topic: "*Alba*" form (p. 60). This western species migrated eastward in the late nineteenth century with the cultivation of alfalfa. First reported in Pennsylvania in 1869.

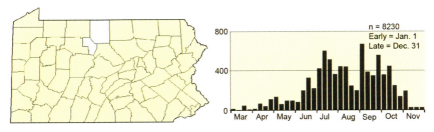

n = 8230
Early = Jan. 1
Late = Dec. 31

800

400

0

Mar   Apr   May   Jun   Jul   Aug   Sep   Oct   Nov

B oth the Clouded Sulphur and Orange Sulphur have a white female form (*"alba"*). By late summer white females make up 20% and 40% of these populations, respectively, and they cannot be reliably distinguished from one another in the field.

How do the opposite sexes of these two species find one another? Females recognize their conspecific males by visual and olfactory cues. Orange Sulphur females, both orange and white forms, recognize males via ultraviolet-reflecting scales on the dorsal surfaces of Orange Sulphur males. These unique scales are absent in Clouded Sulphur males. Theoretically, Clouded Sulphur females are attracted only to males lacking ultraviolet reflectance. Olfactory cues are the final step in male recognition. Males of both species have species-specific pheromones embedded in the cuticles of their wings.

In most instances these sibling species preferentially mate with their own species, yet hybridization does occur throughout the state. Hybrids show an array of variation, making it difficult to assign a specific identification in the field. As with *"alba"* form females, these intermediates should not be classified as one species or another. An interesting topic for investigation would be to quantify the percentage of intermediates in a given area over time. A decreasing percent may signify that isolating mechanisms between the two species are evolving, or one species is swamping out the other.

White *"alba"* forms are commonly found in Pieridae throughout the world. The only other pierid species with a white *"alba"* form commonly found in Pennsylvania is the Little Yellow.

average wing span

Distinguishing marks: Light yellow with pronounced pink wing fringes. Forewing with (1) a small black dot at the end of the discal cell; and reduced black border on female. Ventral hindwing with (2) a single circle at the end of the discal cell. Its distinguishing feature, compared to other *Colias* species in the state, is the lack of markings on the ventral side.

Typical behavior: Frequently nectars. Flies slower than other *Colias* species.

Habitat: Bogs, heaths, and brushy areas with blueberry stands.

Larval hosts: Blueberries (*Vaccinium*).

Abundance: Rare. S1S2

Remarks: One brood. Overwinters as partially grown larva. Known from only two sites in northern Pennsylvania. First discovered in the state in 1968. Can be easily overlooked. Other *Colias* species have faint pink fringes, but those of the Pink-edged Sulphur are thicker and more conspicuous.

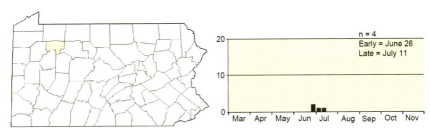

n = 4
Early = June 26
Late = July 11

average wing span

Distinguishing marks: Dorsal forewing with (1) very pointed apex; wide black border; and (2) profile of a dog's face. Ventral hindwing of male with (3) sex patch at base of costa.

Typical behavior: Frequently nectars and is a very fast flier.

Habitat: Any open area, including fields, pastures, gardens, etc.

Larval hosts: False indigo bush (*Amorpha fruticosa*).

Abundance: Rare migrant. SNA

Remarks: Does not breed or overwinter in state. A common species in the South. Occasionally migrates north to Pennsylvania in late summer or fall. Females highly variable in dorsal black markings. Some late fall individuals have prominent rosy scaling (form "*rosa*").

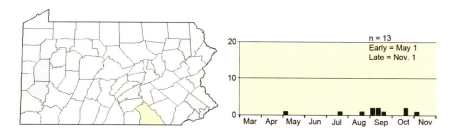

n = 13
Early = May 1
Late = Nov. 1

Cloudless Sulphur   *Phoebis sennae eubule*   (Linnaeus, 1767)

average wing span

Distinguishing marks: Male dorsal side unmarked lemon yellow; female darker yellow with (1) brown scalloped border; and (2) black spot at the end of discal cell on forewing. Male ventral side lightly speckled; female darker yellow with heavier speckling; and (3) silvery spots outlined by brown at end of discal cells.

Typical behavior: Frequently nectars. Strong flier, often flies in a straight beeline path when migrating.

Habitat: Any open areas, including gardens, pastures, meadows, parks, etc.

Larval hosts: Sennas, especially American senna (*Senna hebecarpa*) and Maryland senna (*Senna marilandica*).

Abundance: Uncommon, annual migrant. Rarely breeds. SNA

Remarks: May breed in state upon arrival, producing 1–2 broods in summer and fall. Does not survive winter in state.

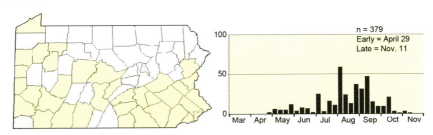

n = 379
Early = April 29
Late = Nov. 11

Orange-barred Sulphur *Phoebis philea philea* (Linnaeus, 1763)

average wing span

Distinguishing marks: Large. Male with (1) diagonal orange bar on the dorsal forewing; and (2) orange patch on border of dorsal hindwing. Female darker with (3) brown scalloped border. Female color varies from pale white to yellow, orange, or red (shown above).

Typical behavior: Very strong flier. Frequently nectars.

Habitat: Any open areas, including gardens, pastures, meadows, and parks.

Larval hosts: Uses large sennas in the South.

Abundance: Extremely rare stray. SNA

Remarks: Does not breed in state or survive winters. Largest sulphur found in Pennsylvania. Collected only once at State College (Centre County) on August 8, 1930. Several records from other northern states occurred in the 1930s.

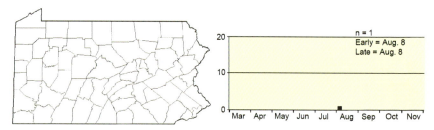

♂   ♀

average wing span

Distinguishing marks: Male dorsal side orange with black borders; female pale orange or yellow with reduced black border on hindwing. Ventral hindwing laced with orange-brown blotches. Ventral side ground color more reddish in fall.

Typical behavior: Low flier with erratic flight. Can move fast when disturbed.

Habitat: Open areas, including fields, meadows, vacant lots, and parks.

Larval hosts: Sennas, especially American senna (*Senna hebecarpa*) and Maryland senna (*Senna marilandica*).

Abundance: Annual migrant from the South, frequently establishes temporary colonies in southern part of the state. S3S4B

Remarks: Breeds in state upon arrival, producing 1–3 broods, but generally does not survive winter. The recorded early April date may indicate that a few adults in reproductive diapause can survive mild winters. The common name is based on the discal cell end bars on dorsal forewings, thought to look like two sleepy eyes.

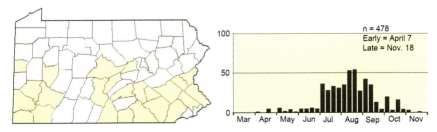

n = 478
Early = April 7
Late = Nov. 18

average wing span

Distinguishing marks: Male dorsal side yellow with black border; female pale yellow to white (form "*alba*") with black border reduced on hindwing. Ventral side with reddish-orange fringe; and (1) orange spot on outer edge of hindwing.

Typical behavior: Low, quick, erratic flight from one flower to next.

Habitat: Open areas, including fields, roadsides, waste areas, and parks.

Larval hosts: Partridge pea (*Chamaecrista fasciculata*) and sensitive partridge pea (*C. nictitans*).

Abundance: Annual migrant from the South, frequently establishes temporary colonies in southern part of the state. S3S4B

Remarks: Breeds in state upon arrival, producing 1–3 broods, but does not survive winter. A surprisingly strong migrator for such a small butterfly. Occasionally reaches northern counties.

Dainty Sulphur   *Nathalis iole iole*   Boisduval, 1836

average wing span

Distinguishing marks: (1) Black bar on inner margin of dorsal forewings; and (2) two black spots on ventral forewing. Male with (3) orange sex patch near costal margin of dorsal hindwing, which fades with age (see magnified view above).

Typical behavior: Very low flier. Rarely flies more than a foot above ground level.

Habitat: Dry open areas, including fields, roadsides, railroad right-of-ways, etc.

Larval hosts: Only recorded hostplant in Pennsylvania is green carpetweed (*Mollugo verticillata*).

Abundance: Rare migrant from southern Mississippi valley and southwestern United States. SNA

Remarks: Can breed in state upon arrival, producing 1–3 broods, but does not survive winter. It is the smallest sulphur found in Pennsylvania. It ventures into the state many years apart. Last seen in large numbers in 2012.

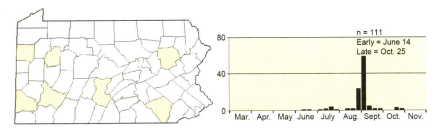

The family Lycaenidae consists of a worldwide group of small, colorful butterflies. Because of their fine, delicate wings, they are commonly referred to as "gossamer wings." Four subfamilies make up the family. These are the harvesters, whose larvae are predatory; the coppers, which are copper-colored; the hairstreaks, which usually bear small hairlike tails on their hindwings; and the blues, which tend to be blue on the dorsal side (particularly males).

Clockwise from the upper left: Harvester, American Copper, Eastern Tailed-Blue, and Gray Hairstreak, from the four subfamilies that make up the Lycaenidae.

Members of the Lycaenidae are the smallest butterflies in the state. They vary in flight and behavior. Harvesters are fast and erratic but never venture far from their aphid hosts. Coppers are less erratic and usually fly slow and low over the ground where their host occurs. Hairstreaks have a fast zigzag flight, making them hard to follow in the air. They are best seen nectaring or perching on leaves near their hosts. Blues are generally weak fliers, staying low to the ground. They are capable of rapid flight when disturbed, and a few species are capable of long-distance migration.

At rest, hairstreaks and blues fold their wings vertically over their bodies and often rub them together. This rubbing behavior is believed to divert strikes of potential predators (e.g., jumping spiders, birds) away from the butterfly's head and toward a less vulnerable "eyespot" on the outer margin of the hindwing. When this area is accompanied by fine hairlike tails ("antennae"), it resembles a false head. Adults are occasionally seen in the field with pieces missing from the "false head," providing evidence of a successful escape from a previous predator attack.

Eggs of this family are flattened and resemble small white turbans. The eggs are laid singly on leaf surfaces or tucked between flowering parts of hostplants. In one subfamily (harvesters) they are placed within aphid colonies. The larvae are generally slug-shaped, and many (particularly hairstreaks and blues) are tended by ants. The ants protect the larvae from parasitic flies and wasps. The chrysalises are small and brown, usually constructed in ground litter or held to small twigs by a silk girdle.

Two examples of the Azure Complex: Summer Azure (left) and Spring Azure (right).

♂   ♀

1

average wing span

Distinguishing marks: Dorsal forewing orange field surrounded by broad black borders; ventral ground color orange-brown with (1) reddish-brown spots encircled in white. Unlike any other lycaenid in North America.

Typical behavior: Does not nectar, but is attracted to dung, sap, and aphid honeydew. Males often found puddling.

Habitat: Woodlands, usually near streams or swamps.

Larval hosts: The only carnivorous butterfly species in North America. Larvae feed on woolly aphids infesting alders, beech, elm, maple, hawthorn, and smilax.

Abundance: Uncommon, local. S3*

Remarks: Three to four overlapping broods. Overwinters as pupa. Lipid composition of larval cuticle mimics that of aphid prey, preventing the attending ants from finding larvae and removing them from aphid colony.

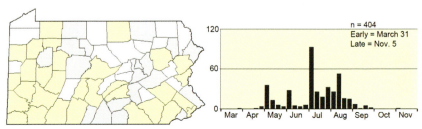

n = 404
Early = March 31
Late = Nov. 5

♂ ♀

average wing span

Distinguishing marks: (1) Coppery orange field in dorsal and ventral forewing with black spots; (2) prominent submarginal orange band on dorsal hindwing; and (3) thin submarginal red-orange line on gray ventral hindwing.

Typical behavior: Flies close to ground. Often found nectaring on small wildflowers. Males territorial, attacking intruders entering their territory.

Habitat: Open areas such as meadows, fields, roadsides, and utility line cuts.

Larval hosts: Sheep sorrel (*Rumex acetosella*).

Abundance: Common, with localized colonies; seemingly absent in some years, common in others. S5

Remarks: Three broods, occasional fourth. Overwinters as partially grown larva. Similar to female Bronze Copper (p. 72), but significantly smaller. Orange submarginal line on ventral hindwing of Bronze Copper is thicker (bandlike). Pockets of population in northeastern portion of state with forewing black spots enlarged and fused (form "*fasciata*").

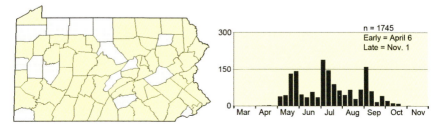

n = 1745
Early = April 6
Late = Nov. 1

♂   ♀

1

1

average wing span

Distinguishing marks: Dorsal forewing of male bronze brown; female forewing with coppery orange field bearing black dots. Hindwing with (1) prominent submarginal orange band on both sides.

Typical behavior: Males perch on low vegetation awaiting females.

Habitat: Open wet meadows, marshes, ditches, and occasionally streamside.

Larval hosts: Various species of dock, including *Rumex crispus*, *R. obtusifolius*, *R. verticillatus*, and *R. orbiculatus*.

Abundance: Uncommon, local. S3*

Remarks: Three broods in southern counties, two in north. Overwinters as egg. Generally forms discrete colonies of a few acres or less. Females may leave wetland habitats and wander several kilometers seeking hostplants. A lone adult encountered away from wetlands is likely to be a dispersing female.

n = 545
Early = May 24
Late = Oct. 20

♂   ♀

average wing span

Distinguishing marks: Dorsum of male with purplish gloss; female nonglossy with many small dark spots. Hindwing with (1) thin submarginal orange line on both sides. Venter pale grayish white.

Typical behavior: Adults commonly nectar on flowers of hostplant.

Habitat: Open cranberry bogs and peaty meadows.

Larval hosts: Cranberry (*Vaccinium macrocarpon* and *V. oxycoccos*).

Abundance: Uncommon, local. S2S3*

Remarks: One brood. Overwinters as egg. This small butterfly is very local and never strays far from bogs where cranberry grows. Males perch on low vegetation and fly out to engage females in courtship. The Bog Copper is found only in northern counties on glaciated plateaus.

n = 661
Early = June 8
Late = July 22

♂   ♀

average wing span

Distinguishing marks: Large hairstreak with two hairlike tails. Dorsal side metallic blue; ventral side black to dark brown; ventral side of abdomen orange.

Typical behavior: Found in late summer nectaring in open fields near hostplant.

Habitat: Woodland clearings and fields near woodlands with mistletoes.

Larval hosts: Mistletoes (family Viscaceae). Uses Christmas mistletoe (*Phoradendron leucarpum*), a parasite of black gum and other trees, south of Pennsylvania.

Abundance: Extremely rare stray in Pennsylvania. SNA

Remarks: A southern hairstreak that presently does not breed in the state. Recorded once in 1999 in Blue Ridge Province (Cumberland County), where historical records of mistletoe exist.

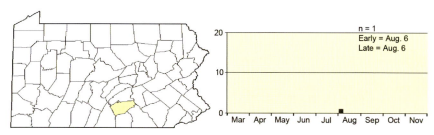

n = 1
Early = Aug. 6
Late = Aug. 6

average wing span

Distinguishing marks: No tails. Ventral hindwing with (1) row of bright red-orange sub-marginal spots; (2) postmedian row of black dots encircled by white; and no blue patch as in other *Satyrium* hairstreaks.

Typical behavior: Adults fond of nectaring on butterfly weed (*Asclepias tuberosa*).

Habitat: Old fields, meadows, forest edges, gardens.

Larval hosts: Black cherry (*Prunus serotina*) and chokecherry (*P. virginiana*).

Abundance: Common, local. S3S4*

Remarks: One brood. Overwinters as egg. Darker individuals lacking white-encircled black dots (ssp. *winteri*) occur in northern glaciated counties. There is some evidence that mature larvae are herded into ant burrows at base of host tree during daytime.

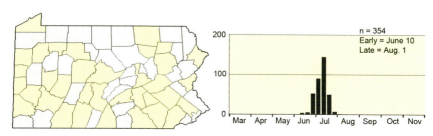

n = 354
Early = June 10
Late = Aug. 1

Acadian Hairstreak *Satyrium acadica acadica* (W. H. Edwards, 1862)

♂ ♀

average wing span

Distinguishing marks: Dorsal hindwing with orange spots near hairlike tail. Venter light gray with (1) postmedian row of round black spots encircled by white on both wings. Hindwing with (2) submarginal row of fused orange spots and (3) blue patch capped with orange.

Typical behavior: Locally found nectaring on milkweeds and dogbane near wetlands.

Habitat: Wetlands, marshes, floodplains with small willows.

Larval hosts: Silky willow (*Salix sericea*) and Bebb willow (*S. bebbiana*).

Abundance: Uncommon, local. S2S3*

Remarks: One brood. Egg overwinters. Mostly found in northern part of state in vicinity of wetlands.

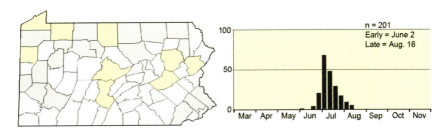

n = 201
Early = June 2
Late = Aug. 16

average wing span

Distinguishing marks: Dorsal side of hindwing with orange spot near hairlike tail. Ventral side light brown with (1) postmedian row of black spots encircled by white on both wings. Ventral hindwing with (2) submarginal row of chevron-shaped orange spots and (3) blue patch near anal angle not capped with orange.

Typical behavior: Found in scrub oak thickets resting on leaves and twigs. Males actively chase one another.

Habitat: Barrens, scrub oak thickets.

Larval hosts: Scrubby oaks, especially bear oak (*Quercus ilicifolia*).

Abundance: Common, local. S3S4*

Remarks: One brood. Egg overwinters. Mostly occurs in eastern part of state. Mature larvae are tended by ants; they feed at night and rest during day in shelters (byres) created by ants at bases of scrubby oaks.

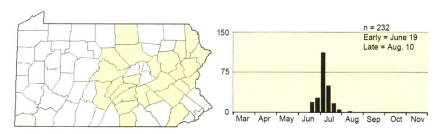

n = 232
Early = June 19
Late = Aug. 10

average wing span

Distinguishing marks: Venter light brown to gray; with (1) postmedian row of dark brown bars bordered by white on one side (male) or both sides (female) on both wings. Hindwing with single prominent orange spot; and (2) blue patch about as wide as long, not capped with orange. (3) Male with oval-shaped stigma.

Typical behavior: Often found nectaring on dogbanes and milkweeds.

Habitat: Woodland edges, clearings, and adjacent fields with nectar flowers.

Larval hosts: Oaks, including white oak (*Quercus alba*), bear oak (*Q. ilicifolia*), and chestnut oak (*Q. prinus*). Also walnuts (*Juglans nigra, J. cinerea*) and pignut hickory (*Carya glabra*).

Abundance: Common. Our most common *Satyrium* hairstreak. S5

Remarks: One brood. Overwinters as egg. Males known for spirally ascending aerial "jousts." Participants usually return to perch with tattered wings.

Hickory Hairstreak *Satyrium caryaevorus* (McDunnough, 1942)

average wing span

Distinguishing marks: Venter light brown with (1) postmedian row of dark brown bars wide and bordered by white on both sides; bars offset and widening near forewing costa; and (2) blue patch on hindwing longer than wide (extended inward). (3) Antennae nudum orange in both sexes (see inset), more intense in female. (4) Male with elliptical stigma, narrower and longer than stigma of Banded Hairstreak (p. 78).

Typical behavior: Most often found nectaring on dogbanes and milkweeds.

Habitat: Woodland edges and clearings, adjacent flowering fields and meadows.

Larval hosts: Bitternut hickory (*Carya cordiformis*) and other hickories. Females observed ovipositing on sprouts of American chestnut (*Castanea dentata*).

Abundance: Uncommon. S3

Remarks: One brood. Overwinters as egg. An uncommon hairstreak easily overlooked. Best identification method uses multiple characters. Worn specimens may need to be examined by an expert.

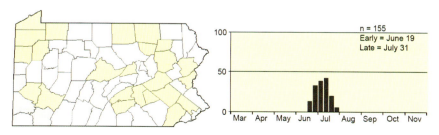

n = 155
Early = June 19
Late = July 31

average wing span

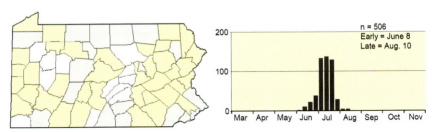

Distinguishing marks: Ventral side light brown with (1) postmedian row of white-bordered bars very wide and strongly offset on both wings; and (2) submarginal blue patch on hind-wing capped with orange.

Typical behavior: Immediately recognized by multiple white lines ("stripes") on ventral side of wings.

Habitat: Woodland clearings and edges.

Larval hosts: Black cherry (*Prunus serotina*) and hawthorns (*Crataegus*).

Abundance: Common, local. S3S4

Remarks: One brood. Overwinters as egg. This species lives primarily in forest canopy. Adults nectar at flowers (especially milkweeds); also found at honeydew droppings on leaves of aphid-infested milkweeds and its host trees.

n = 506
Early = June 8
Late = Aug. 10

average wing span

Distinguishing marks: Dorsal side with small orange spots occasionally present. Dark gray ventral hindwing with no discal markings, (1) postmedian line forming the letter *W* or *M*; and (2) submarginal blue patch on hindwing thinly capped with orange.

Typical behavior: Adults actively nectar, but also seek honeydew droppings.

Habitat: Woodlands, woodland edges, barrens.

Larval hosts: White oak (*Quercus alba*). Likely uses other oaks.

Abundance: Uncommon. Often overlooked. S2S3*

Remarks: One brood. Overwinters as egg. This is a canopy species. Adults imbibe honeydew from wasp galls on woody twigs and from leaves of aphid-infested flowers.

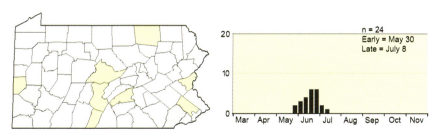

n = 24
Early = May 30
Late = July 8

♂   ♀

average wing span

Distinguishing marks: Dorsal and ventral sides brown. Lacks gray overscaling (frosting) on ventral side as seen in other elfins.

Typical behavior: Found in early spring flying low to ground and puddling on dirt roads and trails.

Habitat: Woodland openings and edges, ridgetop barrens, heaths.

Larval hosts: Mountain laurel (*Kalmia latifolia*) throughout most of state. Blueberries (*Vaccinium angustifolium* and *V. pallidum*) on ridgetop barrens.

Abundance: Common, local. S3S4*

Remarks: One brood. Pupa overwinters. Elfins appear in spring and often prove difficult to identify. Careful attention to ventral hindwing provides accurate identification.

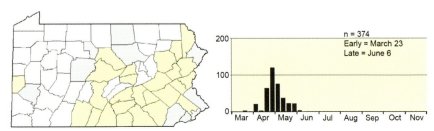

n = 374
Early = March 23
Late = June 6

Hoary Elfin   *Callophrys polios polios*   (Cook and F. Watson, 1907)

♂   ♀

average wing span

Distinguishing marks: Dorsum and venter brown. (1) Extensive gray frosting on outer half of ventral hindwing and small outer portion of forewing. No eyespot on ventral hindwing.

Typical behavior: Small elfin with weak flight, close to the ground.

Habitat: Open woodlands with mats of trailing arbutus.

Larval hosts: Trailing arbutus (*Epigaea repens*).

Abundance: Very rare. Possibly extirpated. SH*

Remarks: One brood. Pupa overwinters. Smallest elfin in state; not recorded since 1985.

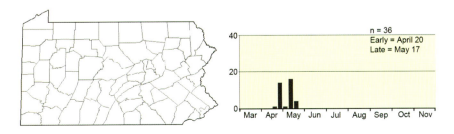

n = 36
Early = April 20
Late = May 17

average wing span

Distinguishing marks: Dorsal and ventral sides brown. Very short tail on hindwing. Ventral hindwing with (1) gray frosting on outer half, containing small black eyespot. (2) Male with thin elliptical stigma.

Typical behavior: Flies in open trails and power line cuts close to hostplants. Infrequently at flowers.

Habitat: Woodland openings and edges, barrens, sandy soils, power line cuts.

Larval hosts: Wild indigo (*Baptisia tinctoria*) and wild lupine (*Lupinus perennis*).

Abundance: Uncommon to rare, local. S1S2*

Remarks: One brood. Pupa overwinters. Wild indigo ecotype is most common in state. Individuals that feed on wild indigo tend to be larger than lupine feeders. Both are declining in the northeastern United States.

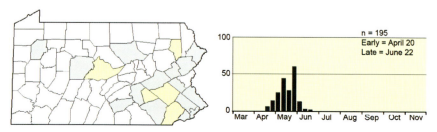

Henry's Elfin *Callophrys henrici henrici* (Grote and Robinson, 1867)

♂ ♀

average wing span

Distinguishing marks: Short tail on hindwing. Ventral hindwing with light frosting on outer half, dark brown on inner half; (1) white postmedian line; and (2) squarelike outward projection. In male, white line reduced to (3) small accents near costal and anal margins. Male lacks stigma.

Typical behavior: Never far from redbud. Males frequently puddle.

Habitat: Woodland clearing, trails, and edges.

Larval hosts: Eastern redbud (*Cercis canadensis*) and wild plum (*Prunus americana*).

Abundance: Uncommon. S2S3*

Remarks: One brood. Pupa overwinters. Sometimes locally common in Ridge and Valley Province.

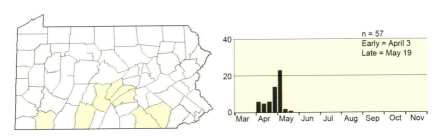

n = 57
Early = April 3
Late = May 19

♂ ♀

average wing span

Distinguishing marks: Fringes heavily checkered. Ventral side boldly patterned with brown and white accents.

Typical behavior: Largest elfin in state. Commonly seen puddling on forest trails.

Habitat: Pine-oak forest clearings and edges.

Larval hosts: Eastern white pine (*Pinus strobus*), pitch pine (*P. rigida*), and Virginia pine (*P. virginiana*).

Abundance: Common, local. S4

Remarks: One brood. Pupa overwinters. Adults prefer to perch in higher branches. Descend to nectar and puddle (males) on trails. Females oviposit on small pines. One of very few butterflies whose larval hostplants are conifers.

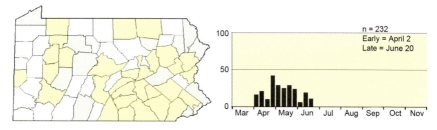

n = 232
Early = April 2
Late = June 20

Juniper Hairstreak *Callophrys gryneus gryneus* (Hübner, [1819])

♂   ♀

average wing span

Distinguishing marks: Single hairlike tail. Ventral side green with white-lined reddish-brown lines and black eyespot near base of tail.

Typical behavior: Males perch high in cedar trees. Shaking trees disturbs them from perch; after a quick flight around tree, they resettle on perch.

Habitat: Forest edges and fields with red cedar. Occasionally on ridgetops.

Larval hosts: Eastern red cedar (*Juniperus virginiana*).

Abundance: Uncommon. In some years, locally common. S3S4*

Remarks: Two broods, occasional partial third brood. Pupa overwinter. Summer brood adults often found nectaring. Only other hairstreak with green on the ventral side is the Early Hairstreak (p. 91), which has a paler green and lacks tails.

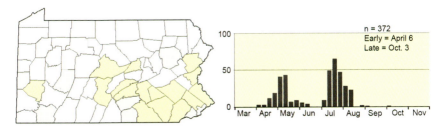

n = 372
Early = April 6
Late = Oct. 3

average wing span

Distinguishing marks: Two tails. Dorsal side brilliant blue with wide black margins. Dark gray ventral hindwing with (1) white dash near midpoint of costa; single red spot inside submarginal line; and (2) postmedian line forming a *W* or *M*.

Typical behavior: Flashing blue iridescence in flight. Actively nectars.

Habitat: Woodland clearings, adjacent meadows and gardens.

Larval hosts: Eastern oaks, including white oak (*Quercus alba*).

Abundance: Uncommon, but increasing numbers in last two decades. S3S4

Remarks: Three broods. Pupa overwinters. Species spreading northward in recent decades.

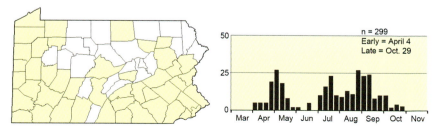

n = 299
Early = April 4
Late = Oct. 29

Gray Hairstreak   *Strymon melinus humuli*   (T. Harris, 1841)

♂   ♀

average wing span

Distinguishing marks: Dorsal side with (1) orange dot at base of tail. Ventral hindwing gray with (2) thin postmedian black line bordered by orange and white; orange patch containing black spot at base of tail; no discal markings.

Typical behavior: Adults fond of garden mint nectar. Larvae occasional pest of garden beans.

Habitat: Woodland edges, meadows, fields, and gardens.

Larval hosts: Wide variety of legumes including tick-trefoils (*Desmodium*), bush clovers (*Lespedeza*), sennas (*Senna*), and string beans; also invasive loosestrife (*Lythrum*).

Abundance: Common throughout state. S5

Remarks: Three broods, possible fourth brood in some years. Pupa overwinters. Commonly found in late summer.

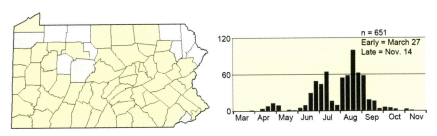

n = 651
Early = March 27
Late = Nov. 14

♂   ♀

average wing span

Distinguishing marks: Two tails. Dorsal side dark gray and male; female with slight blue scaling. Ventral side dark gray with (1) prominent postmedian red band on both wings.

Typical behavior: Flies close to ground. Late summer adults fond of goldenrod nectar.

Habitat: Fields, meadows, woodland edges.

Larval hosts: Decaying leaves and detritus on ground, particularly around sumacs.

Abundance: Common migrant in most years. S4B

Remarks: Multiple broods possible (2–3). A southern species recently expanding its range northward. Breeding typically begins each year upon arrival from the south. Rarely over-winters as partially grown larva.

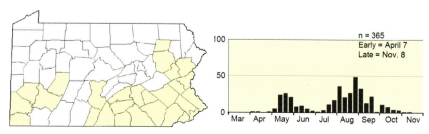

n = 365
Early = April 7
Late = Nov. 8

average wing span

Distinguishing marks: Dorsal side dark gray; female with blue scaling. Ventral side pale green; both wings decorated with orange spots.

Typical behavior: Axiom—"It occurs in the most unlikely place when least expected."

Habitat: Mature forest and clearings, such as roads and trails, with beech trees.

Larval hosts: American beech (*Fagus grandifolia*) and beaked hazelnut (*Corylus cornuta*). Also uses white oak (*Quercus alba*) in mid-Appalachians.

Abundance: Rare. S2*

Remarks: Two broods. Pupa overwinters. This is a canopy species. Both sexes descend to nectar. Males puddle at damp ground. Recorded less than two dozen times in the state. It is a real prize to see this hairstreak.

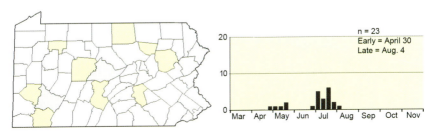

n = 23
Early = April 30
Late = Aug. 4

♂   ♀

average wing span

Distinguishing marks: Dorsal side purplish blue. Ventral side with numerous brown and white lines extending through both wings; two submarginal black eyespots on hindwing.

Typical behavior: Only blue in our region, with alternating brown and white stripes on ventral side.

Habitat: Resident of Southwest, where it occurs in many habitats

Larval hosts: Pea family (legumes) elsewhere. Possibly uses alfalfa (*Medicago sativa*) or white sweet clover (*Melilotus alba*) locally.

Abundance: Rare migrant. SNA

Remarks: One of few lycaenids that perform long-distance migration. Does not overwinter in state. Potentially could breed temporarily. Last appeared in state in 1993 (twice) and 2002.

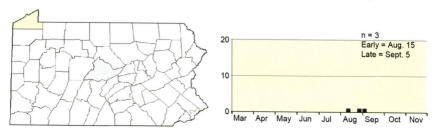

n = 3
Early = Aug. 15
Late = Sept. 5

Eastern Tailed-Blue *Cupido comyntas comyntas* (Godart, [1824])

♂   ♀

average wing span

Distinguishing marks: Small tail on hindwing. Dorsal side blue in male; black in female. Ventral side light gray with small dots and bars; submarginal orange and black spots near base of tail.

Typical behavior: Flies close to the ground. Males puddle.

Habitat: Ubiquitous in dry open areas, fields, roadsides, disturbed areas, and lawns.

Larval hosts: Pea family (legumes), especially clovers (*Trifolium*), bush clovers (*Lespedeza*), tick-trefoils (*Desmodium*), sweet clovers (*Melilotus*), alfalfa (*Medicago sativa*), and crown vetch (*Securigera* [= *Coronilla*] *varia*).

Abundance: Abundant throughout the state. S5

Remarks: Multiple broods (4–5) from spring to fall. Overwinters as fully grown larva. Our only blue with hindwing tail.

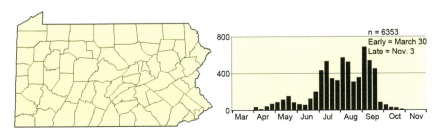

n = 6353
Early = March 30
Late = Nov. 3

form "*marginata*" form "*violacea*"

average wing span

Distinguishing marks: Dorsal side blue to purplish blue; male forewing with subtle white matte surface (see Special Topic: Azure Complex, p. 100); female blue, occasionally with slight purplish tinge, black forewing border, and submarginal dots on hindwing. Wing fringes checkered. Ventral side light or dark gray with variable spot pattern.

Typical behavior: Small blue butterfly flying in spring deciduous woodlands. Males puddle.

Habitat: Deciduous woods with clearing, trails, edges, and service roads.

Larval hosts: Floral buds of flowering dogwood (*Cornus florida*), viburnums (*Viburnum prunifolium, V. acerifolium*) and black cherry (*Prunus serotina*).

Abundance: Common in woodlands in southern part of state, except ridgetops. S5

Remarks: One brood. Pupa overwinters. Variable ventral hindwing patterns. Form "*violacea*" with dots only; "*marginata*" with dark wing margin; and "*lucia*" with coalesced dots in median area (see p. 97). These forms also occur in other *Celastrina* species.

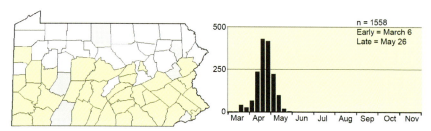

n = 1558
Early = March 6
Late = May 26

500

250

0

Mar Apr May Jun Jul Aug Sep Oct Nov

♂ ♀

average wing span

Distinguishing marks: Dorsal side brilliant sky blue; female with black forewing border and submarginal spots on hindwing. Occasional female with greenish tinge. Wing fringes checkered. Ventral side light or dark gray with variable spot pattern.

Typical behavior: Small blue butterfly flying in northern woodlands during spring. Males puddle.

Habitat: Bogs, heaths, sandy barrens, dry ridgetops, and open northern woodlands.

Larval hosts: Floral bud of blueberries (*Vaccinium corymbosum, V. angustifolium, V. pallidum*) and chokecherry (*Prunus virginiana*).

Abundance: Common in heaths and woodlands in northern part of state. S5

Remarks: One brood. Pupa overwinters. A northern species, common on glaciated plateaus; extends southward on ridgetops in Ridge and Valley Province.

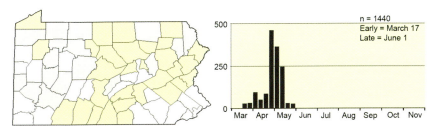

n = 1440
Early = March 17
Late = June 1

♂ ♀

average wing span

Distinguishing marks: Dorsal side light blue with white scaling between veins, especially on hindwing; female with black forewing border; and submarginal spots on hindwing. Fringes on hindwing not checkered. Ventral side white with greatly reduced spots.

Typical behavior: This is the only azure flying in summer and fall.

Habitat: Fields, parks, gardens, roadsides, forest edges, and streamside.

Larval hosts: Floral buds of many plant families, particularly shrubby dogwoods (*Cornus racemosa*, *C. amomum*), ninebark (*Physocarpus opulifolius*), white meadowsweet (*Spiraea alba*), steeplebush (*Spiraea tomentosa*), hog peanut (*Amphicarpaea bracteata*), horsebalm (*Collinsonia canadensis*), and wingstem (*Verbesina* [=*Actinomeris*] *alternifolia*).

Abundance: Common, found throughout the state. S5

Remarks: Multiple overlapping broods (3–4), rare partial fifth brood. This is the only Azure flying in summer and fall. Overwinters as pupa. The uncommon spring form (April) has little white scaling on dorsal side, and ventral side is bright white with larger black spots.

n = 6331
Early = April 5
Late = Nov. 5

Cherry Gall Azure   *Celastrina serotina*   Pavulaan and D. Wright, 2005

♂

♀

form *"lucia"*

average wing span

$\longleftarrow \qquad \longrightarrow$

Distinguishing marks: Dorsal side blue; some individuals with slight white scaling between hindwing veins; female with black forewing border and submarginal spots on hindwing. Fringes minimally checkered. Ventral side light gray to white with variable spot pattern.

Typical behavior: Avid nectarer. Males puddle. Fresh individuals appear as flight of Northern Spring Azure (p. 95) is waning. Open northern woodlands, sandy barrens, and dry ridgetops.

Larval hosts: Stalked galls on leaves of black cherry (*Prunus serotina*) and chokecherry (*P. virginiana*). Floral buds of nannyberry (*Viburnum lentago*), Appalachian tea (*V. cassinoides*), alternate-leaf dogwood (*Cornus alternifolia*), and red osier dogwood (*C. sericea* [*=stolonifera*]).

Abundance: Moderately common. Paired with Northern Spring Azure in sympatry. S4

Remarks: One brood. Pupa overwinters. Long recognized as a distinct flight between flights of spring and summer species in northern habitats.

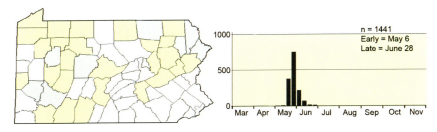

n = 1441
Early = May 6
Late = June 28

♂ ♀

average wing span

Distinguishing marks: Dorsal side blue; some individuals with slight white scaling between veins; female with black forewing border and submarginal spots on hindwing. Fringes minimally checkered. Ventral side white with reduced black spots; some individuals without spots (immaculate).

Typical behavior: Avid nectarer. Males puddle. Large fresh individuals appear as flight of Spring Azure is waning.

Habitat: Similar to Spring Azure (p. 94). Deciduous woods with clearings where black cohosh grows.

Larval hosts: Floral buds of black cohosh (*Actaea* [= *Cimicifuga*] *racemosa*).

Abundance: Common, local. Paired with Spring Azure in sympatry. S3S4*

Remarks: One brood. Pupa overwinters. Long recognized as a distinct flight between flights of spring and summer species in southern and central Appalachians.

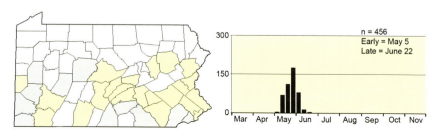

n = 456
Early = May 5
Late = June 22

♂   ♀

average wing span

Distinguishing marks: Dorsum black in male; blue in female with wide black border and faint submarginal spots on hindwing. Fringes checkered. Venter gray with numerous black spots and dashes.

Typical behavior: Small black butterfly flying with Spring Azure (p. 94). Males puddle.

Habitat: Shaded moist deciduous woodlands, north-facing slopes, streamsides, and roadsides where goatsbeard grows.

Larval hosts: Young leaves and flowers of goatsbeard (*Aruncus dioicus*).

Abundance: Rare. Possibly extirpated. SH*

Remarks: One brood. Pupa overwinters. Discovered by W. H. Edwards in West Virginia in the nineteenth century when it was believed to be a dimorphic male of Spring Azure. Elevated to species level in 1960. Occurs in sympatry with Spring Azure and Appalachian Azure. Last recorded in state in 1968. Surveys of habitat and hostplant needed.

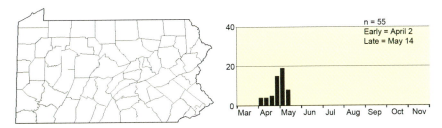

n = 55
Early = April 2
Late = May 14

Six closely related species make up the Azure Complex in Pennsylvania. As a group they show a bewildering array of colors, marks, host preferences, and flight periods. At times they were collectively considered to be a single, continentwide, variable species. In addition, individuals of the spring flight of the Spring Azure (*C. ladon*, above right) were thought to produce the lighter-colored individuals of the summer flights (*C. neglecta*, above left). This paradigm shifted when it was discovered that Spring Azures fly just once a year and do not produce a second brood. Further investigation established that the scale morphology of summer *C. neglecta* males is very different from that of spring *C. ladon* males. Under magnification (below) the dorsal forewing of *C. neglecta* (left) can be seen to consist of alternating rows of blue scales and small white scales. The latter are modified scales known as *androconia*, which bear pheromones. *C. ladon* (below right) stands in stark contrast. *C. ladon* wings lack androconia and instead feature a long overlapping translucent scale of unknown function. Scanning electron micrographs (SEMs) provide more details of the different scale patterns.

All magnified photos are of right dorsal forewing, with base toward the left.

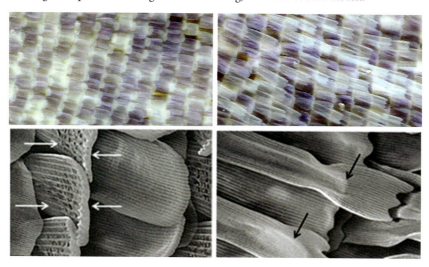

Species of the Azure Complex are floral specialists and generally lay their eggs on unopened floral buds. Each species routinely couples their adult eclosion (*ex* pupa) to coincide with the floral budding period of local favorite hostplants. This allochronic pattern provides their larvae with a rich supply of nutrients for fast growth and development. It also explains how Azure species have diversified. The chart below depicts when peak oviposition (marked by ■) occurs on the floral host of each species in Pennsylvania.

Males of each species are shown below and are at life size.

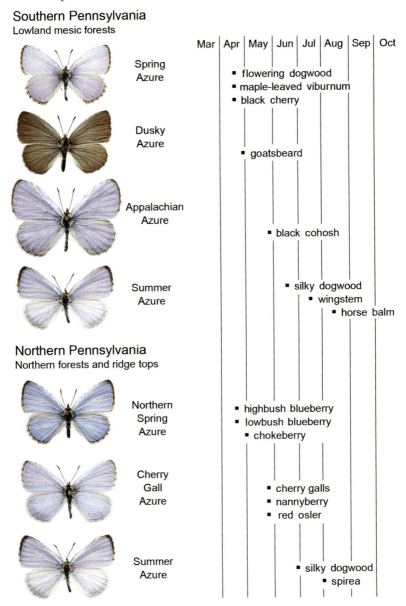

Southern Pennsylvania
Lowland mesic forests

	Mar	Apr	May	Jun	Jul	Aug	Sep	Oct

Spring Azure
- ■ flowering dogwood
- ■ maple-leaved viburnum
- ■ black cherry

Dusky Azure
- ■ goatsbeard

Appalachian Azure
- ■ black cohosh

Summer Azure
- ■ silky dogwood
- ■ wingstem
- ■ horse balm

Northern Pennsylvania
Northern forests and ridge tops

Northern Spring Azure
- ■ highbush blueberry
- ■ lowbush blueberry
- ■ chokeberry

Cherry Gall Azure
- ■ cherry galls
- ■ nannyberry
- ■ red osler

Summer Azure
- ■ silky dogwood
- ■ spirea

average wing span

Distinguishing marks: Dorsal side bright silvery blue; female with black borders. Ventral side dark gray with row of large submarginal black spots encircled by white on both wings.

Typical behavior: Very small blue butterfly of spring woodlands, flying low to the ground. Males puddle. Often flies with Spring Azure (p. 94).

Habitat: Moist, deciduous Appalachian forests with host vetch.

Larval hosts: Carolina or wood vetch (*Vicia caroliniana*).

Abundance: Uncommon to rare. Declining statewide. S1S2*

Remarks: One brood. Pupa overwinters. *G. l. couperi* Grote, 1873, a larger subspecies with smaller black dots, strong flight, and a different host, cow vetch (*V. cracca*), began moving southward from Canada in recent decades. There are two records of this subspecies straying briefly in Pennsylvania in 1964 and 1969. *G. l. lygdamus* (E. Doubleday, 1841), an even larger subspecies, occurs in open pine forests on the sandy coastal plain in the southeastern United States.

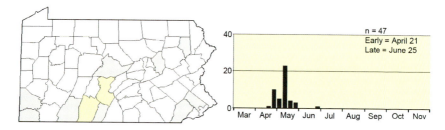

n = 47
Early = April 21
Late = June 25

Karner Blue — *Plebejus samuelis* — (Nabakov, 1944)

♂ ♀

average wing span

Distinguishing marks: Dorsal side dark blue in male; black in female with slight blue scaling and submarginal orange band on hindwing. Ventral side gray-white with submarginal orange band on hindwing, punctuated with iridescent blue spots (scintillae). It is the only blue with both ventral wings margined with orange.

Typical behavior: Avid nectarer. Males puddle.

Habitat: Acid sandy soils with wild lupine (*Lupinus perennis*), such as pine and oak savannas and pine barrens.

Larval hosts: Wild lupine.

Abundance: Rare. Presumed extirpated. SX*

Remarks: Two broods. Egg overwinters. Originally discovered in Karner, New York, in 1869. First reported in Pennsylvania in 1874 and last collected in 1977, as a dispersing female. The nineteenth-century record likely from summer brood, no specific dates available.

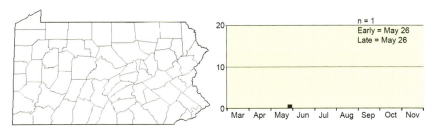

n = 1
Early = May 26
Late = May 26

The family Riodinidae is a vast family of small to medium-sized butterflies centered principally in the Neotropics, especially tropical America. Their common name comes from small metallic spots on the wings of many species (see Special Topic: Metalmarks, p. 105). There are over 1,500 species in this family, of which 29 have been recorded in the United States and only two in Pennsylvania. The latter are considered rare in the state. Metalmarks are closely related to the Lycaenidae and some taxonomists previously treated them as a subfamily of Lycaenidae.

Individuals in this family are weak fliers and they generally tend to stay in the vicinity of their larval hostplants. While brilliant tones and colors characterize the tropical species, the North American species are subtly colored gray or rusty brown. At rest they perch with wings spread flat; when disturbed, some perch on the underside of leaves. On sunny days they are avid nectarers.

The hostplants of metalmark larvae represent more than forty different plant families. The Pennsylvania species feed on a woodland ragwort and a wetland thistle, members of Asteraceae. The eggs are shaped like turbans. The larvae actively feed on leaves and are plump and heavily decorated with long thin hairs, giving them a distinctive hairy appearance. Most riodinid larvae in the Neotropics have evolved close associations with ants. They have specialized organs that act to soothe ants in return for protection from predatory wasps and flies, which the ants chase away. It is not known if our Pennsylvania metalmarks are associated with ants. Much research remains to be performed. The larvae overwinter. Pupation occurs at the base of the hostplant or in nearby leaf litter.

Our metalmarks reside in habitats (shale barrens, swamps) that are prone to risks of drainage, clearcutting, and development. Preservation of these critical habitats would help save local metalmark populations, as well as preserve flora and fauna that share their realm.

A mating pair of Northern Metalmarks: male (left) and female (right).

Ventral side of male Northern Metalmark.

Special Topic   Metalmarks

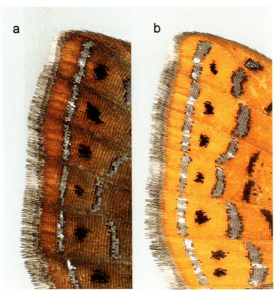

As mentioned above, members of the family Riodinidae are commonly known as metalmarks due to the metallic markings on both dorsal and ventral wing surfaces. While this can be seen to some extent in the species accounts that follow, it is worthwhile to have magnified views where the metallic nature of the markings is more easily seen.

The photos on the left show the outer portion of the male forewing of a Northern Metalmark, (a) dorsal side and (b) ventral side.

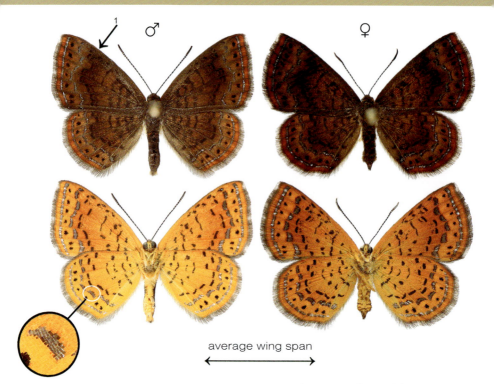

average wing span

Distinguishing marks: Dorsal side orange and brown (male lighter) with (1) faint dark band running through center of wings and two rows of silver spots on both wings. Ventral side bright orange with postmedian row of crescent-shaped silvery spots on ventral hindwing (see inset).

Typical behavior: Flies weakly, close to the ground. Perches with wings spread. Occasionally lands on underside of leaves.

Habitat: Openings in forests or woodland, especially on limestone or shale ridges. Also in power line cuts in same areas.

Larval hosts: Round-leaved ragwort (*Packera* [=*Senecio*] *obovata*).

Abundance: Uncommon to rare, with localized colonies. S1S2*

Remarks: One brood. Overwinters as partially grown larva. Most colonies today persist in Ridge and Valley Province.

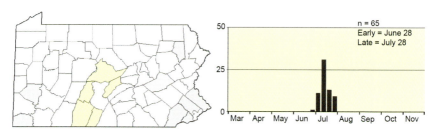

n = 65
Early = June 28
Late = July 28

average wing span

Distinguishing marks: Similar to Northern Metalmark, but smaller. (1) Lacks dark median band on dorsal wings; postmedian row of silvery spots on ventral hindwing more rectangular than crescent-shaped (see inset).

Typical behavior: Flies weakly, close to the ground. Never far from thistles. Occasionally lands on underside of leaves.

Habitat: Wooded areas along streams and moist soggy areas near fens.

Larval hosts: Unknown in Pennsylvania, generally swamp thistle (*Cirsium muticum*) and roadside thistle (*C. altissimum*).

Abundance: Rare. Presumed extirpated. SX

Remarks: One brood. Overwinters as partially grown larva. Recorded only once in western Pennsylvania, in the 1930s. This occurrence appears to be a remnant of a very widely scattered and fragmented population in the Ohio River valley.

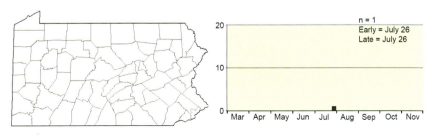

n = 1
Early = July 26
Late = July 26

The family Nymphalidae consists of a diverse group of small to large butterflies found worldwide, commonly known as the brushfoots. The family gets its name from its small front legs, which are often so small that they are not easily recognizable. The front legs are not engaged in walking or standing. Instead, the incomplete front pair of feet are supplied with numerous small hairs resembling a brush, which the butterfly uses to smell and taste.

In eastern North America the family consists of seven subfamilies, including the snouts, milkweed butterflies, longwings and fritillaries, true brushfoots, admirals, emperors, and satyrs and browns. Members of the family are highly variable in color and patterns. Many are brown and camouflaged on both sides of their wings. Others are brown on the underside while brightly colored on the topside. Some have rounded wings with smooth margins, while others have irregular notched margins resembling leaves. A few have brightly colored wings and body, indicating that they are either toxic or mimics of other species that are toxic, and thus avoid predators.

Hackberry Emperor (left) and Monarch (right).

Variegated Fritillary (left) and Viceroy (right).

Red-spotted Purple (left) and Mourning Cloak (right).

Common Wood Nymph subspecies nephele *(left) and subspecies* alope *(right). Both are found in Pennsylvania.*

Brushfoot adults are strong fliers and are quite noticeable when active in daylight. They can be found in almost any habitat that has forests or patches of weeds. Though they frequently sip nectar, many species in this family prefer tree sap, rotting fruit, dung, or decomposing animal carcasses. Males seek mates by either perching or patrolling, and they attract mates with courtship displays and scent pheromones. After mating, females lay up to several hundred eggs, often leaving a scent mark on plants where they have laid eggs to deter other females from doing the same. Some species lay eggs in clusters, others in columns or singly.

Brushfoot larvae eat many different kinds of plants. Some specialize on thistles, nettles, willows, milkweeds, daisies, violets, etc., and a few specialize on grasses and sedges. Many of the caterpillars have bumps on their bodies and projecting horns or spines to discourage predators. Some are dark-colored; others are green or yellow; many have stripes or spots. The chrysalids often mimic twigs or decaying vegetation. Usually it is the larvae that hibernate in this family, but a few species hibernate as adults.

Some brushfoot adults are among the longest-lived butterflies in North America, surviving just short of a year (10–11 months), such as the Compton Tortoiseshell and Mourning Cloak. Some adults migrate long distances annually to reach Pennsylvania in late spring or summer, such as the Monarch, Painted Lady, Red Admiral, and Common Buckeye.

From top left clockwise: Atlantis Fritillary, Aphrodite Fritillary, Diana Fritillary, and Great Spangled Fritillary. The Great Spangled, Aphrodite, and Atlantis Fritillaries are the most common large fritillaries in Pennsylvania. Note the gray-blue eye color of the Atlantis Fritillary (top left). This character helps identify this species in the field; other large fritillaries have orangeish-brown eyes. The unique eye color of the Atlantis Fritillary is lost in museum specimens.

American Snout *Libytheana carinenta bachmanii* (Kirtland, 1851)

average wing span

Distinguishing marks: Dorsal side orange and black. (1) Labial palps very long, giving appearance of a snout or beak; (2) forewing apex hooked, similar to anglewings; and (3) subapical white spots on forewing.

Typical behavior: Adults often rest on vegetation or tree trunks with head facing down.

Habitat: Open woods, woodland edges, roadsides, and streamsides where hackberry grows.

Larval hosts: Hackberries (*Celtis occidentalis, C. tenuifolia*). Rarely, hop vine (*Humulus lupulus*).

Abundance: Common, local. S3S4B

Remarks: Three to four broods depending on when adults arrive in state. A southern migrant that generally cannot survive winters in Pennsylvania.

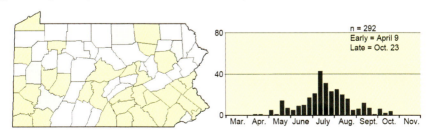

n = 292
Early = April 9
Late = Oct. 23

average wing span

Distinguishing marks: Large orange butterfly with black veins and black borders containing rows of white dots. Dorsal forewing with (1) subapical yellowish-orange spots. Male hindwing with (2) black patch of scent scales.

Typical behavior: Slow, deliberate flight. Avid nectarer. In September and October, large numbers observed in south-southwest-directed flight. During fall migration, groups rest overnight in trees.

Habitat: Open areas, meadows, old fields, roadsides, power line right-of-ways, gardens.

Larval hosts: Milkweeds (*Asclepias syriaca, A. incarnata, A. tuberosa, A. curassavica*).

♀

Abundance: Common. Apparently vulnerable due to decline in overwintering forest habitat in Mexico, modified agricultural practices in United States, and poor management of native milkweed patches. S3S4B

Remarks: Two to four broods in state depending on arrival of northbound spring migrants. A well-known migrating butterfly that accumulates cardiac glycoside toxins in larval stages; adults serve as distasteful model in mimicry complex.

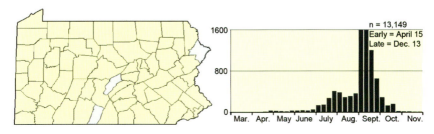

n = 13,149
Early = April 15
Late = Dec. 13

average wing span

Distinguishing marks: Similar to Monarch (p. 112), but distinguished by dark reddish-brown color, lack of black veins on dorsum, and lack of black transverse markings in forewing apex. Males with (1) black patch of scent scales on dorsal hindwing. Looks like a dark Monarch.

Typical behavior: Avid nectarer.

Habitat: Open areas, old fields, meadows, roadsides, and gardens.

Larval hosts: Milkweeds and milkweed vines in the South.

♀

Abundance: Rare migrant. SNA

Remarks: Year-round resident of the deep South. Does not breed in Pennsylvania. Sightings in Pennsylvania occurred in 2002 and 2003, when there were sudden appearances of this species in the Northeast.

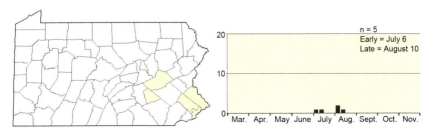

n = 5
Early = July 6
Late = August 10

20

10

0

Mar. Apr. May June July Aug. Sept. Oct. Nov.

♂   ♀

average wing span

Distinguishing marks: Dorsal side bright orange. Elongated forewings with pointed apex. Ventral side with large silvery spots on hindwing.

Typical behavior: Avid nectarer. Fast, strong flier.

Habitat: Open spaces, old fields, meadows, pastures, and gardens.

Larval hosts: Passion vines (*Passiflora*) in the South.

Abundance: Rare migrant. SNA

Remarks: Year-round resident of the deep South. Does not breed in Pennsylvania. Last recorded in state in 1961.

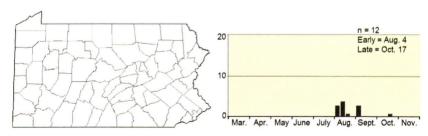

n = 12
Early = Aug. 4
Late = Oct. 17

Variegated Fritillary   *Euptoieta claudia claudia*   (Cramer, 1775)

♂   ♀

average wing span

Distinguishing marks: Dorsal side golden-orange with zigzag black markings. Ventral side mottled with variable shades of brown, tan, and gray. Basal area of ventral hindwing with white veins.

Typical behavior: Fast flier, difficult to catch or photograph except when nectaring at flowers.

Habitat: Open areas, old fields, meadows, pastures, and gardens.

Larval hosts: Violets (*Viola*) and passion vines (*Passiflora*).

Abundance: Increasingly common migrant in late summer and fall. S3S4B

Remarks: One to three broods depending on when adults arrive. Overwinters from Virginia southward. Cannot survive winters in Pennsylvania. The numbers recorded have increased in the last two decades.

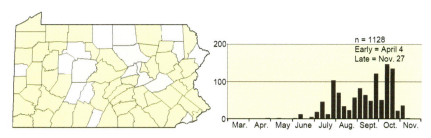

n = 1128
Early = April 4
Late = Nov. 27

average wing span

Distinguishing marks: Dorsal side two-toned; dark brown near body. Forewing (1) lacking inner (postbasal) black dot present in other large fritillaries. Ventral hindwing inwardly brown with (2) wide yellow submarginal band between two rows of silvery spots ("spangled").

Typical behavior: Avid nectarer. Strong flier. Wanders widely.

Habitat: Open areas, old fields, wet meadows, roadsides, and gardens.

Larval hosts: Common blue violet (*Viola sororia*) and other violets. Females often oviposit on grass or dried leaves adjacent to violets.

Abundance: Our most common large fritillary. Common throughout the state. S5

Remarks: One brood. Overwinters as unfed first instar larva. Males emerge before females in early part of flight season. Mating occurs in midsummer. Thereafter, females go into reproductive diapause and reappear in August.

Five large fritillaries (genus *Speyeria*) have been recorded in Pennsylvania. Only three are commonly seen in the state. Of the remaining pair, one has been recorded only once and the other, common a century ago, is now restricted to a single colony at Fort Indiantown Gap. See introduction to nymphalids for photographs of specimens in their environment.

The three most common large fritillaries are very similar. All are orange with the black markings forming spots, chevrons, and zigzag lines. Each species has its own distinguishing marks.

♀

n = 6349
Early = May 13
Late = Oct. 17

1400

700

0
Mar Apr May Jun Jul Aug Sep Oct Nov

average wing span

Distinguishing marks: Dorsal side orange with little brown next to body. Forewing with (1) inner black dot. Ventral hindwing inwardly reddish brown. Reddish brown color surrounds large silvery white spots and (2) extends into yellow submarginal band, narrowing the band.

Typical behavior: An avid nectarer. Flies quicker than the Great Spangled Fritillary (p. 118).

Habitat: Upland wooded areas, grassy fields, dry meadows. Occasionally bogs.

Larval hosts: Common blue violet (*Viola sororia*) and other violets.

Abundance: Common, but less so than Great Spangled Fritillary. Tends to prefer higher elevations in northern part of state. S3S4

♀

Remarks: One brood. Overwinters as unfed first instar larva. Males emerge before females. On average Aphrodite Fritillary begins flight a few days later than Great Spangled Fritillary. Easily overlooked when mingling with the former.

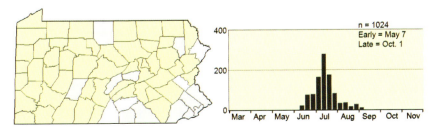

n = 1024
Early = May 7
Late = Oct. 1

average wing span

Distinguishing marks: Dorsum orange with little brown next to body. Forewing with (1) inner black dot; and (2) thick black marginal lines (best seen in female). Ventral hindwing inwardly dark brown mixed with gray. Dark color (3) extends into yellow submarginal band, appearing as heavy dots at tips of silvery spots. Live individuals have grayish-blue eye color.

Typical behavior: Avid nectarer. Attracted to animal droppings (scat).

Habitat: Cool, open areas of northern forest and adjacent meadows, streamsides, and glades

Larval hosts: Violets (*Viola*). Precise violet species not reported in Pennsylvania.

Abundance: Uncommon, except for northern counties. S3*

♀

2

Remarks: One brood. Overwinters as unfed first instar larva. Smallest of the three common large fritillaries. More likely to be found in cooler woodlands of northern part of state and Laurel Highlands. Occasionally strays into state from the North.

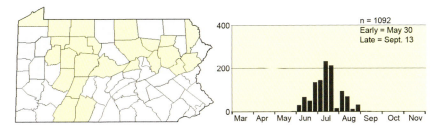

n = 1092
Early = May 30
Late = Sept. 13

♂

average wing span

Distinguishing marks: Dorsal side with reddish-orange forewing and bluish-black hindwing. Hindwing with rows of white and orange spots (male) or white spots only (female). Ventral hindwing chocolate brown with numerous silvery triangular spots and lacking submarginal yellow band. The only fritillary with reddish-orange forewing and black hindwing.

Typical behavior: Flight is lower and more gliding than other large fritillaries.

Habitat: Variable grasslands, including damp meadows, mountain pastures, grassy prairies, old fields. Occurs in local restricted colonies.

Larval hosts: Arrowleaf violet (*Viola sagittata*), perhaps other *Viola* species in the past.

Abundance: Rare, locally restricted. Critically imperiled in state. S1*

♀

Remarks: One brood. Overwinters as unfed first instar larva. The Regal Fritillary formerly occupied a large portion of the Northeast to southern Appalachians including all of Pennsylvania. Rapid decline began post–World War II following loss of habitat to agriculture and urbanization. Use of herbicides may have also contributed to decline. A single protected colony persists at Fort Indiantown Gap.

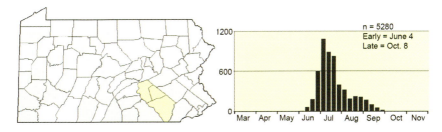

n = 5280
Early = June 4
Late = Oct. 8

♂

average wing span

Distinguishing marks: Dorsal side of male brown-black with outer third orange. Female black with bluish-white spots on forewing and iridescent blue on hindwing. Ventral side of male orange, female black. Stunning example of sexual dimorphism.

Typical behavior: Bluish-black females reclusive; retreat into deep woods when disturbed.

Habitat: Moist woodlands with rich soil. Found along shady woodland roads, trails, and adjacent meadows.

Larval hosts: Violets (*Viola*).

Abundance: Rare. Presumed extirpated. SX

♀

Remarks: One brood. Overwinters as unfed first instar larva. A southern butterfly, recorded only once in late nineteenth century as part of historical population in Ohio River drainage. Female is a mimic of distasteful Pipevine Swallowtail (p. 30).

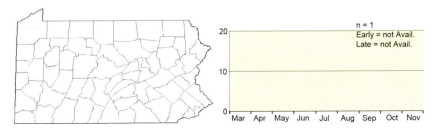

n = 1
Early = not Avail.
Late = not Avail.

♂   ♀

average wing span

Distinguishing marks: Dorsal side similar to large fritillaries, but much smaller. Ventral side reddish orange with rows of metallic silver spots.

Typical behavior: Avid nectarer. Rarely strays from moist, open habitat. Flight is low and erratic.

Habitat: Wet meadows, marshes, bogs, moist old fields.

Larval hosts: Violets (*Viola*). Precise species not reported in Pennsylvania.

Abundance: Uncommon. Most often encountered in higher elevations of northern counties in localized colonies. S2S3*

Remarks: Two broods, possible partial third brood in southern counties. Overwinters as partially grown larva. Usually co-occurs with the more common Meadow Fritillary (p. 129), with which it can be confused. Both are referred to as lesser fritillaries.

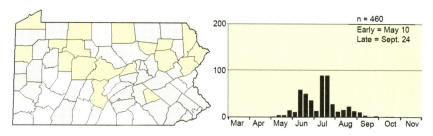

n = 460
Early = May 10
Late = Sept. 24

Meadow Fritillary   *Boloria bellona bellona*   (Fabricius, 1775)

♂   ♀

average wing span

Distinguishing marks: Dorsal side similar to large fritillaries, but much smaller. Forewing apex long and squared-off. Ventral hindwing variegated with brown, gray, purple, and yellowish orange. Outer margin dusted in purple with row of faint spots. No silvery spots.

Typical behavior: Low-flying and seldom settles for more than a brief period of time.

Habitat: Moist meadows, old fields, roadsides, marshes, bogs.

Larval hosts: Violets (*Viola*). Precise species not reported in Pennsylvania. Eggs laid on other plants in vicinity of violets.

Abundance: Common throughout the state. S5

Remarks: Three complete broods with partial fourth brood in some years. Overwinters as partially grown larva. This species can be expected to occur wherever moist, open habitats exist, including urban areas.

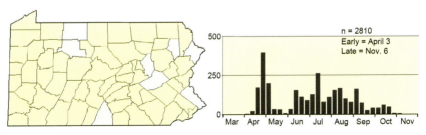

n = 2810
Early = April 3
Late = Nov. 6

The following four species are checkerspots. The first three species (genus *Chlosyne*) have similar dorsal sides; close inspection of ventral hindwings is needed for proper identification. The fourth species, Baltimore Checkerspot (genus *Euphydryas*), is significantly larger, with considerable black on the dorsal side.

♂   ♀

1→

average wing span

Distinguishing marks: Dorsal hindwing with (1) pale orange chevrons in black marginal band. Ventral hindwing with contrasting zigzag white bars interspersed with brown lines.

Typical behavior: Avid nectarer. Males puddle at moist locations.

Habitat: Old grassy fields, meadows, and streamsides.

Larval hosts: Precise hosts unknown in Pennsylvania.

Abundance: Rare. Presumed extirpated in early twentieth century. SX*

Remarks: Probably two broods in historical colony in Lackawanna County. Last collected in June 1906. Isolated populations in the East are relicts of postglacial Xerothermic period, which extended the range of prairie species eastward.

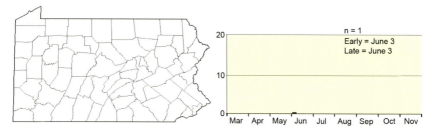

n = 1
Early = June 3
Late = June 3

♂ ♀

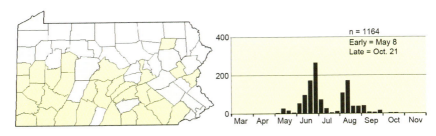

average wing span

Distinguishing marks: Resembles Pearl Crescent (p. 134), but is much larger. Dorsal hindwing with (1) black spots within submarginal orange band, some spots having orange centers. Ventral hindwing creamy white highlighted with brown veins; and (2) one white crescent in middle of brown patch on margin.

Typical behavior: Slow, weak flier, staying close to the ground.

Habitat: Openings in damp woods, streamsides, meadows, roadsides.

Larval hosts: Members of Asteraceae, including wingstem (*Verbesina* [=*Actinomeris*] *alternifolia*), cutleaf coneflower (*Rudbeckia laciniata*), Jerusalem artichoke (*Helianthus tuberosus*), New York ironweed (*Vernonia noveboracensis*), flat-topped aster (*Doellingeria umbellata*), and purple stem aster (*Symphyotrichum puniceum*).

Abundance: Uncommon, local. More common in southern counties. S3S4*

Remarks: Two broods. Overwinters as partially grown larva. Larvae are gregarious and may strip host.

n = 1164
Early = May 8
Late = Oct. 21

average wing span

Distinguishing marks: Dorsal side darker than previous species. Dorsal hindwing with (1) black spots within submarginal orange band touching marginal band. Ventral hindwing with orange and white bands divided by black lines; and (2) row of prominent white crescents along margin.

Typical behavior: Slow flier, staying close to the ground. Males often at damp soil.

Habitat: Moist meadows, marshes, roadsides.

Larval hosts: Flat-topped aster (*Doellingeria umbellata*).

Abundance: Uncommon. Found most often in isolated populations in damp meadows in northern counties. S3*

Remarks: One brood. Overwinters as partially grown larva at base of hostplant. Adult and larva are possible mimics of the distasteful Baltimore Checkerspot (p. 133).

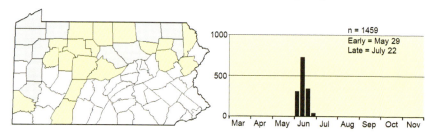

n = 1459
Early = May 29
Late = July 22

♂ ♀

average wing span

Distinguishing marks: Distinctive black butterfly with numerous orange and white spots on both sides. Female forewing with rounded contour and reduced orange bars in discal area (often none at all).

Typical behavior: Perches on low vegetation in bright sun with wings spread.

Habitat: Wet meadows, marshes, edges of bogs, streamsides. Populations using narrowleaf plantain found in dry open fields.

Larval hosts: Multiple hosts in several plant families. White turtlehead (*Chelone glabra*), yellow false foxglove (*Aureolaria flava*), Canadian lousewort (*Pedicularis canadensis*), narrowleaf plantain (*Plantago lanceolata*), and others.

Abundance: Uncommon. Found in isolated populations. S3*

Remarks: One brood. Overwinters as partially grown larva. Both adults and larvae are distasteful to avian predators. Black with bright orange pattern is considered to be a warning (aposematic).

n = 1119
Early = May 19
Late = Aug. 10

Crescents (genus *Phyciodes*) form a large complex of species found continentwide. The number of sibling species and taxonomy are still being worked out. There are three species in Pennsylvania, distinguished by their size, timing of broods, orange patch on dorsal hindwing, and antennal club shape and color (see Special Topic: Antennal Clubs, p. 135).

summer form "*morpheus*"

fall/spring form "*marcia*"

average wing span

Distinguishing marks: Dorsal hindwing with (1) orange patch divided by black veins. Ventral hindwing with (2) silvery crescent in middle of brown patch on margin. Antennal club round to oval with underside orange (female) or black/orange (male). There are two seasonal forms: (a) summer form "*morpheus*" with light creamy yellow ground color on venter, and (b) darker fall/spring form "*marcia*" with varying shades of gray, brown, white, and yellow.

Typical behavior: Avid nectarer. Flies low over vegetation. Pugnacious.

Habitat: Meadows, pastures, weedy fields, vacant urban lots.

Larval hosts: Numerous asters, including purple stem aster (*Symphyotrichum puniceum*), hairy aster (*S. pilosum*), calico aster (*S. lateriflorum*), and white panicle aster (*S. lanceolatum*).

Abundance: Abundant throughout the state. S5

Remarks: Three broods, with fourth brood in many years. Overwinters as partially grown larva. Starts flight earlier in spring than other crescents. Rare two-brooded local colonies with orange antenna males occur in the state. They may represent a relict *tharos* type or a different species closely related to the next species.

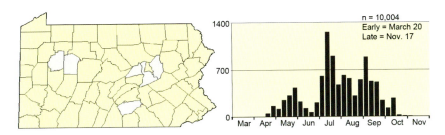

n = 10,004
Early = March 20
Late = Nov. 17

Special Topic — Antennal Clubs

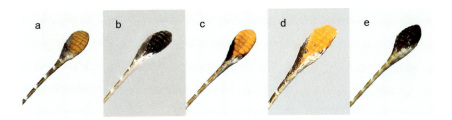

The color of the scaleless area of the underside of the antennal club, known as the nudum, can be helpful in identifying crescents, particularly males. Shown below are the lower surfaces of antennal clubs of: (a) female *tharos*, (b) male *tharos*, standard black color; (c) male *tharos* rare orange color; (d) male *cocyta*; and (e) male *batesii*.

♂ ♀

average wing span

Distinguishing marks: Similar to Pearl Crescent (p. 134), but larger. Dorsal hindwing with (1) orange patch open and undivided by black veins. Ventral hindwing with (2) silvery crescent in middle of small brown patch on margin. Antennal club elongate with underside orange in both sexes.

Typical behavior: Flies high over vegetation in flap-and-glide style. When disturbed, flies onto lower tree branches.

Habitat: Open rocky places and barrens, plus moist areas near streams, meadows, and marshes.

Larval hosts: White panicle aster (*Symphyotrichum lanceolatum*) and crooked stem aster (*S. prenanthoides*).

Abundance: Uncommon. Found in local restricted colonies in northern counties. S3*

Remarks: One brood. Overwinters as partially grown larva. Flies between first and second brood of Pearl Crescent. A second brood is currently unknown. Placement of this entity as subspecies *selenis* is tentative. The taxonomy of the *cocyta* group is currently under study.

n = 30
Early = June 5
Late = July 21

♂   ♀

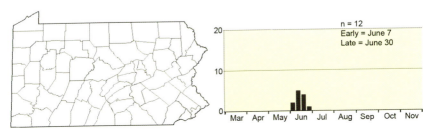

average wing span

Distinguishing marks: Similar to Pearl Crescent (p. 134), but darker. Dorsal hindwing with (1) orange patch small and divided by black veins. Ventral hindwing with (2) faint silvery crescent and absent brown patch. Antennal club round to oval with underside black in both sexes. When flying, appears like a dark-colored Pearl Crescent. When perched with wings folded, the pure yellow hindwing is evident.

Typical behavior:

Habitat: Open rocky places, barrens, plus moist areas near streams and marshes.

Larval hosts: Wavyleaf aster (*Symphyotrichum undulatum*). Possibly other asters.

Abundance: Rare. Presumed extirpated in late 1960s. SX*

Remarks: One brood with short season. Overwinters as partially grown larva. In the early twentieth century this species was not uncommon. Last collected in state in 1965.

n = 12
Early = June 7
Late = June 30

20

10

0

Mar Apr May Jun Jul Aug Sep Oct Nov

Common Buckeye   *Junonia coenia coenia*   Hübner, [1822]

♂   ♀

average wing span

Distinguishing marks: Immediately recognizable by three large dorsal eyespots, a creamy white band enveloping forewing eyespot, and two small orange bands on forewing.

Typical behavior: Active flier and nectarer. Males sit on bare ground awaiting females.

Habitat: Dry open areas. Often found on bare ground under power line cuts and beside railroad tracks.

Larval hosts: Narrowleaf plantain (*Plantago lanceolata*), purple gerardia (*Agalinis purpurea*), and blue toadflax (*Nuttallanthus canadensis*).

Abundance: A common migrant from the south in most years, arriving midsummer to fall. Rarely appears in spring. S5B

n = 1743
Early = March 30
Late = Dec. 13

form "*rosa*"

Remarks: Summer individuals have light clay-colored ventral hindwings. Most individuals emerging in the fall have dark reddish-tan ventral hindwings; some have a deep rose color (form "*rosa*") shown above (left). Does not survive winters in Pennsylvania.

Special Topic   Brushfoots

Members of one butterfly family (Nymphalidae) appear to have two pairs of legs rather than three pairs. The front pair is greatly reduced in size and held close to the body; these legs are not used for walking. Females use the front legs to sample potential host plants for possible oviposition.

In photo (a) on the right, all three pairs of legs are shown in the "*rosa*" form of the Common Buckeye; the left leg of the small front pair is indicated by an arrow. In (b) a further magnification shows only the front pair of legs.

The photos show the dense, hairy nature of these forelegs, giving rise to the name "brushfoots."

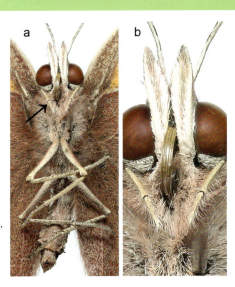

The following four species are known collectively as anglewings, based on their angular wing shapes. These four are similar in appearance, with their ventral surfaces resembling dead leaves or tree bark. They are similar in (1) habitat selection, being mainly woodland species; (2) nutrient choices, preferring rotting fruit, tree sap, carrion, dung, and damp soil; (3) sexual dimorphism primarily evident on the ventral side of their wings; and (4) seasonal dimorphism with a summer form and a fall/spring form (except Green Comma). Adults of the fall/spring form hibernate in tree crevices, hollow logs, and similar refuges over the winter.

Magnified views of the punctuation marks (question mark, comma) for which these species are named are shown in Special Topics: Punctuation Marks.

summer form "*umbrosa*"

average wing span

Distinguishing marks: Dorsal side orange and black. Ventral side resembles dead leaf. Seasonal forms. Both forms with (1) an extra spot on dorsal forewing not present in other anglewings; (2) longer tail than other anglewings; and (3) silvery white question mark in the middle of ventral hindwing. Summer form (shown above) with very dark dorsal hindwing. Fall/spring form (on next page) with lighter dorsal hindwing, visible orange spots, and violet scaling of tail.

♂ ♀

fall/spring (or winter) form "*fabricii*"

Typical behavior: Rarely nectars. Attracted to sap, rotting fruits, carrion, and dung. Often found puddling or basking in afternoon sun.

Habitat: Woodland openings and edges.

Larval hosts: Hackberry (*Celtis*), elms (*Ulmus*), nettle (*Urtica*), and Japanese hop (*Humulus japonicus*).

Abundance: Common throughout the state. Short-distance migrant. Breeds in state. S5

Remarks: Two overlapping broods. Overwinters as adult. In fall, migrates south of Mason-Dixon Line (or to southern counties) prior to hibernation. In spring, northward migrants repopulate the state. Returning adults sometimes appear in irruptive numbers.

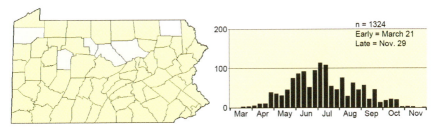

n = 1324
Early = March 21
Late = Nov. 29

♂   ♀

summer form "*dryas*"

average wing span

Distinguishing marks: Similar to Question Mark (p. 000). Seasonal forms. Both forms (1) lacking extra spot on dorsal forewing, which is present in Question Mark; (2) a short tail; and (3) a silvery white comma in the middle of ventral hindwing. Summer form (shown above) with dark dorsal hindwing. Fall/spring form (on next page) with lighter dorsal hindwing visible orange spots, and slight violet scaling of tail.

Typical behavior: Rarely nectars, but is attracted to sap, rotting fruits, and dung. Males often puddle.

Habitat: Woodland openings and edges.

Larval hosts: Nettles (*Urtica*), elms (*Ulmus*), hackberry (*Celtis*), and Japanese hop (*Humulus japonicus*).

Abundance: Common throughout the state. S5

Remarks: Two broods. Overwinters as adult. Nonmigratory, or minimally so. Compare flight phenograms of the Eastern Comma and Question Mark (p. 141). The fall/spring form appears very early in spring.

fall/spring (or winter) form *"comma"*

n = 1856
Early = Feb. 11
Late = Dec. 9

Special Topic   Punctuation Marks

The exact shape and thickness of the silvery white commas and question marks found on anglewings can often be used to differentiate between two species. Six examples are shown below: (a) and (b) Question Mark with question marks varying in thickness and proportions; (c) and (d) Eastern Comma with variable hooked ends; (e) Green Comma with hooked ends; and (f) Gray Comma with typical thin comma, lacking hooked ends.

♂   ♀

average wing span

Distinguishing marks: Similar to previous anglewings. Ventral side resembles lichen-encrusted tree bark. No seasonal forms. Wing margins (1) very ragged; and (2) green chevrons in submargin on venter (see magnified view).

Typical behavior: Rarely nectars. Attracted to sap and rotting fruit. Males often puddle.

Habitat: Woodland openings and edges.

Larval hosts: Birches (*Betula*), willows (*Salix*), and alders (*Alnus*).

Abundance: Rare. Recently recorded in Forest County following a long period during which none were recorded. S1*

Remarks: One brood. Overwinters as adult. A generally northern species that should be looked for in the northern regions of the state.

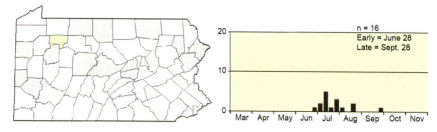

n = 16
Early = June 28
Late = Sept. 28

Mar | Apr | May | Jun | Jul | Aug | Sep | Oct | Nov

average wing span

Distinguishing marks: Slightly smaller than other anglewings. Seasonal forms. Both forms with ragged wing margins. Dorsal hindwing (1) lacking black spot found in other anglewings. Ventral wings strongly two-toned, gray and dark brown with vertical striae; and hindwing with (2) thin silvery white comma lacking hook (see Special Topic: Punctuation Marks, p. 143). Fall/spring form is shown above. Summer form (*"l-argenteum"*) has darker dorsal hindwing.

Typical behavior: Rarely nectars, but is attracted to sap, rotting fruits, and dung. Males often puddle.

Habitat: Woodland openings and edges.

Larval hosts: Restricted to currants (*Ribes*).

Abundance: Uncommon, but not rare. S3S4.

Remarks: Two broods. Overwinters as adult. Lower arm of white comma is shorter than upper arm, thus appearing more like letter *L* than a comma.

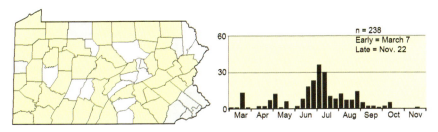

n = 238
Early = March 7
Late = Nov. 22

Shown above, from top left clockwise, are ventral sides of Gorgone, Silvery, Baltimore, and Harris' Checkerspots.

Shown below, from left to right, are two examples of dorsal sides of Pearl Crescents (male and female).

The ventral sides of Pearl Crescents are quite variable, as seen in the three examples shown above. From left to right are summer female, spring female, and summer male.

Shown below are four anglewings. The top row illustrates dorsal views of the Question Mark (left) and the Eastern Comma (right). Note the extra black spot on the forewing of the Question Mark. The bottom row illustrates ventral views of the Gray Comma (left) and the Eastern Comma (right). Note the thin, unhooked comma of the Gray and the thicker, hooked comma of the Eastern.

average wing span

Distinguishing marks: Ventral side mimics tree bark or dead leaf. Dorsal side dark with (1) pale white spots on both wings. Ventral hindwing marbled gray with (2) small white mark resembling an *L* or *J* at the end of the cell.

Typical behavior: Rarely nectars, but is attracted to sap and rotting fruits, and males puddle.

Habitat: Woodland openings and edges.

Larval hosts: Elms (*Ulmus*), willows (*Salix*), birches (*Betula*), and aspens (*Populus*).

Abundance: Uncommon. S3*

Remarks: One brood. Overwinters as adult. Spring individuals produce fresh adults in early summer, which ultimately hibernate through upcoming winter. In some years, population enhanced by migrants from the north.

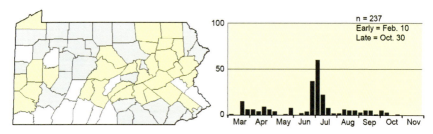

♂   ♀

average wing span

Distinguishing marks: Similar to Compton Tortoiseshell (p. 148), but smaller. Dorsal hindwing with extensive orange. Venter marbled brown with green spots in submargin.

Typical behavior: Rarely nectars. Attracted to sap and rotting fruit. Males puddle.

Habitat: Woodland openings and edges, and stream corridors.

Larval hosts: Does not breed in state. Buckbrushes (*Ceanothus*) in the West.

Abundance: Very rare migrant. Recorded once over seventy years ago. SNA

Remarks: The California Tortoiseshell is found primarily in the West. In outbreak years it is known to travel great distances eastward. Three individuals were taken in Pennsylvania at bait in early September 1945.

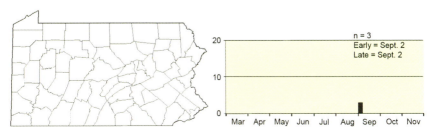

n = 3
Early = Sept. 2
Late = Sept. 2

20

10

0

Mar Apr May Jun Jul Aug Sep Oct Nov

average wing span

Distinguishing marks: Dorsal side maroon-brown with yellow borders; ventral side black with buff borders. Nothing similar found in Pennsylvania.

Typical behavior: Rarely nectars. Attracted to sap and rotting fruit. Males puddle.

Habitat: Woodlands, forest edges, parks, river courses, and city yards.

Larval hosts: Many broadleaf trees, including elms (*Ulmus*), willows (*Salix*), and Japanese hop (*Humulus japonicus*).

Abundance: Common throughout the state. S5

Remarks: One brood. Overwinters as adult. A familar butterfly, often one of the first observed in February or March. Spring individuals produce fresh adults in early summer; they are infrequently seen after mid-July, as they estivate to preserve body fat. Short southward migrations in fall have been observed.

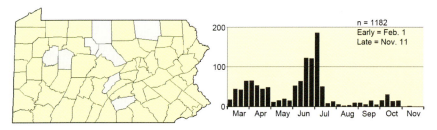

♂   ♀

average wing span

Distinguishing marks: Dorsum two-toned; dark chocolate-brown inner half and flaming orange outer half. Orange bar at end of cell on forewing. Marked like no other butterfly in Pennsylvania.

Typical behavior: Nectars avidly. Also fond of sap, rotting fruit, and dung.

Habitat: Woodland edges, moist meadows, streamsides, and rockslides.

Larval hosts: Stinging nettle (*Urtica dioica*).

Abundance: Uncommon. S3S4

Remarks: Two broods, possible third brood in some years. Overwinters as adult. The main portion of its range lies north of our area. Species has periodic outbreaks, suddenly appearing in large numbers. Adults occasionally migrate into state from the North during the fall.

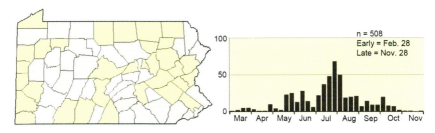

n = 508
Early = Feb. 28
Late = Nov. 28

♂ ♀

average wing span

Distinguishing marks: Dorsal side orange and brown-black. Forewing with (1) small white dot, sometimes missing in male. Ventral hindwing with (2) two large eyes.

Typical behavior: Avid nectarer. Favors low vegetation with nectar flowers.

Habitat: Open areas like meadows, parks, gardens, roadsides, and forest edges.

Larval hosts: Several hosts, primarily pearly everlasting (*Anaphalis margaritacea*) and field pussytoes (*Antennaria neglecta*).

Abundance: Common migrant. Appears earlier in state than Painted Lady (p. 153). S5B

Remarks: Most years does not survive winter in state. Overwinters as adult south of Mason-Dixon Line. Spring migrants annually repopulate the state. Three broods, rarely a fourth brood.

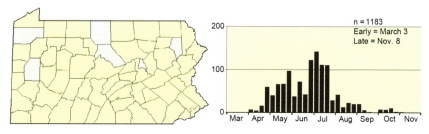

n = 1183
Early = March 3
Late = Nov. 8

average wing span

Distinguishing marks: Similar to American Lady (p. 152). Forewing lacks white dot. Ventral hindwing with (1) five small eyes.

Typical behavior: Avid nectarer. Favors low vegetation with nectar flowers. More skittish than American Lady.

Habitat: Open areas like meadows, parks, gardens, roadsides, and forest edges.

Larval hosts: Thistles (*Cirsium, Carduus, Echinops*), borage (*Borago officinalis*), and common mallow (*Malva neglecta*).

Abundance: Common migrant. Numbers vary widely year to year. S5B

Remarks: One to three broods depending on when migrants populate the state. Does not survive most winters in Pennsylvania. Overwinters as adult in South and Southwest. Typically arrives in late summer.

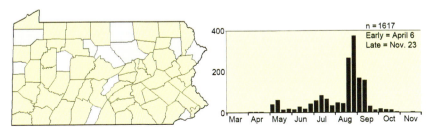

average wing span

Distinguishing marks: Dorsal side black with red-orange band across forewing and along hindwing border.

Typical behavior: Avid nectarer. Also attracted to dung and sap.

Habitat: Any open areas like meadows, roadsides, parks, and gardens.

Larval hosts: Various nettles (*Urtica, Laportea, Boehmeria*).

Abundance: Common migrant. Sometimes appears in massive numbers. S5B

Remarks: Does not survive most winters in Pennsylvania. Two to three broods depending on when migrants populate the state. Typically arrives in spring. Massive spring migrations can occur. See Special Topics: Irruptions (p. 162).

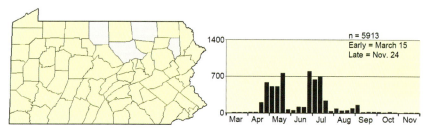

n = 5913
Early = March 15
Late = Nov. 24

The genus name *Vanessa* means "butterfly" in Greek. Members of this group are commonly called Vanessas. There are three Vanessas found in Pennsylvania.

Painted Lady (left) and American Lady (right), note white dot on forewing. See Distinguishing marks (p. 152).

Red Admiral.

There are two species of the genus *Limenitis* in Pennsylvania. Both are effective mimics of other distasteful butterflies. The Viceroy mimics the Monarch. The Red-spotted Purple (part of a polytypic species) mimics the Pipevine Swallowtail.

average wing span

Distinguishing marks: Dorsal side black with white band across both wings. Ventral side with brick-red spots basally and in submarginal rows.

Typical behavior: Males puddle. Also fond of sap, rotting fruit, and dung. When disturbed, apt to fly considerable distance and land in trees.

Habitat: Open northern woodlands, trails, forest roads, and streamsides.

Larval hosts: Quaking aspen (*Populus tremuloides*).

Abundance: Common. More common in northern counties. S5

Remarks: Two overlapping broods. Overwinters as partially grown larva. Pennsylvania resides in broad blend zone where this subspecies and the following subspecies hybridize. Individuals with traces of white band occur in this zone.

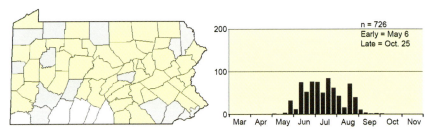

n = 726
Early = May 6
Late = Oct. 25

Red-spotted Purple *Limenitis arthemis astyanax* (Fabricius, 1775)

average wing span

Distinguishing marks: Dorsal side with iridescent blue on hindwing. Mimics Pipevine Swallowtail (p. 30). Ventral side with brick-red spots basally and in submarginal rows. On occasion, dorsal hindwing may be iridescent green (form "*viridis*").

Typical behavior: Commonly found mineralizing on moist gravel or dirt roads. Also fond of sap, rotting fruit, and dung. When disturbed, apt to fly short distance and return.

Habitat: Open mesic woodlands, trails, forest roads, and streamsides.

Larval hosts: Black cherry (*Prunus serotina*), quaking aspen (*Populus tremuloides*).

Abundance: Common throughout the state. S5

Remarks: Two broods, rare partial third brood. Overwinters as partially grown larva. Predominates in southern portion of blend zone where Pipevine Swallowtail is most common. May interbreed with Viceroy (p. 159), producing unusual hybrid "*rubidus*".

n = 3835
Early = April 15
Late = Oct. 25

Pennsylvania is located geographically (latitude 39°43′ N to 42° N) where many northern and southern butterflies meet. This situation permits opportunities to explore areas of contact. These areas are commonly known as hybrid or blend zones. Within such zones, individuals may hybridize, creating intergrades that have features intermediate between the parental forms.

The northern and southern subspecies of *Limenitis arthemis* meet in a zone stretching along a line running east–west through the middle of the state. In this hybrid zone individuals of the fully banded northern subspecies *arthemis* (White Admiral) mate with nonbanded individuals of the southern subspecies *astyanax* (Red-spotted Purple). The result is intergrade individuals with partial white bands. Examples of the dorsal sides of two such intergrades are shown below (top row).

Likewise the northern and southern subspecies of the Common Wood Nymph (*Cercyonis pegala*) meet along a line running east–west through the middle of the state. In this hybrid zone individuals of southern subspecies *alope* (eyespots in yellow patch) mate with individuals of northern subspecies *nephele* (eyespots in brown patch). Examples of the dorsal and ventral sides of two intergrades are shown below (bottom rows). Note the intermediate degree of yellow shading around the forewing eyespots.

form "*albofasciata*"   form "*proserpina*"

♂   ♀

average wing span

Distinguishing marks: Similar to the Monarch, but smaller. Hindwing with (1) black post-median line.

Typical behavior: Generally seen nectaring. Males are very territorial and attack anything, including humans entering their territory.

Habitat: Edge of moist woods or meadows with streams or ponds.

Larval hosts: Generally willows (*Salix*) but also poplars (*Populus*).

Abundance: Uncommon, but not rare. S4

Remarks: Two broods, rare partial third brood. Overwinters as partially grown larva; third instar larva constructs a hibernaculum of host leaves and silk.

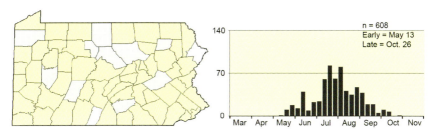

n = 608
Early = May 13
Late = Oct. 26

average wing span

Distinguishing marks: Gray and brown. Forewing with (1) black eyespot. Hindwing with submarginal row of distinct eyespots; and (2) small black spot on inner margin of ventral hindwing, variable in size.

Typical behavior: Drawn to dung, sap, carrion, and rotting fruit. Often perches head-down on sides of tree trunks.

Habitat: Open woodlands, floodplains, and parks.

Larval hosts: Hackberries, usually common hackberry (*Celtis occidentalis*) and dwarf hackberry (*C. tenuifolia*).

Abundance: Common. S4

Remarks: Two broods. Overwinters as partially grown larva. Males smaller than females and their forewings are more falcate. Generally found flying around hackberry trees.

n = 664
Early = May 23
Late = Oct. 7

average wing span

Distinguishing marks: Tawny orangeish brown. Similar to Hackberry Emperor (p. 160). Forewing with (1) eyespot lacking. Hindwing with submarginal row of distinct eyespots; and (2) small black spot on inner margin on ventral hindwing lacking.

Typical behavior: Drawn to dung, sap, carrion, and rotting fruit. Generally found flying around hackberry trees.

Habitat: Open woodlands, floodplains, and parks.

Larval hosts: Hackberries, usually common hackberry (*Celtis occidentalis*) and dwarf hackberry (*C. tenuifolia*).

Abundance: Uncommon, but not rare. S3S4

Remarks: One brood. Overwinters as partially grown larva. Males noticeably smaller than females and their forewings are more falcate.

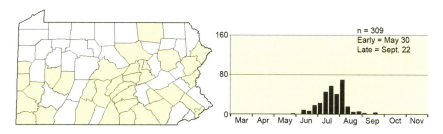

n = 309
Early = May 30
Late = Sept. 22

There is a great deal of variation in the coloring of the hindwings of this species, mainly tending toward darker forms. In an extremely dark form ("*proserpina*"), all markings on the hindwing are nearly obscured.

form "*proserpina*"

Special Topic   Irruptions

In zoology an *irruption* refers to a sudden dramatic increase in population density, which in the butterfly world most often involves migrants. A striking irruption of northbound Red Admirals occurred in Pennsylvania and surrounding states during early spring of 2012. Shown below are two flight phenograms of the Red Admiral. The one on the left includes data from 2012 and clearly delineates a sudden surge of individuals in April. The one on the right shows all flight data (excluding the 2012 irruption) and reflects a typical year with two to three standard broods.

A question arises as to what happened to the large influx of individuals in April 2012. Was there a large subsequent brood in June and July? The answer is not so good from the butterfly's perspective. In one locality (Beaver County), extremely large numbers of eggs were laid on local nettles. Within one to two weeks after emerging, the larvae had completely defoliated every nettle plant in the area. They literally ate themselves out of house and home; their remains littered the ground. Only a small number made it to the adulthood, as evidenced by no later spikes in the flight phenogram.

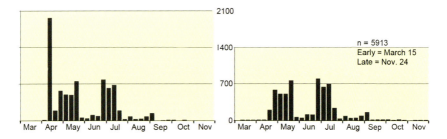

average wing span

Distinguishing marks: Dorsal side light brown with (1) eyespots larger than the browns (next two species) and lacking central white dot (pupil). (2) Hindwing margin prominently scalloped. Ventral side with whitish cast on outer half of forewing; and hindwing with violet cast (when fresh).

Typical behavior: Males perch upside down on tree trunks. Attracted to tree sap, moist soil, and dung.

Habitat: Deciduous woodlands with sparse understory and large stands of grass.

Larval hosts: Woodland grasses, particularly bearded shorthusk (*Brachyelytrum erectum*) and bottlebrush grass (*Hystrix patula*).

Abundance: Common. Usually found in local colonies. S4

Remarks: One brood. Overwinters as partially grown larva. Appears to be partially bivoltine in extreme southern counties.

n = 1316
Early = May 6
Late = Oct. 2

average wing span

Distinguishing marks: Dorsal side hindwing with (1) terminal eyespots bearing central white dot (pupil). Ventral side light brown with (2) jagged dark postmedian line on both wings, separating dark and light areas.

Typical behavior: Favors open, sunny wetlands.

Habitat: Open sedge meadows, marshes, and fens.

Larval hosts: Various sedges (*Carex* species). *Carex lacustris, C. stricta, C. rostrata,* and *C. trichocarpa* accepted in the lab.

Abundance: Uncommon. Occurs in localized colonies. S2S3*

Remarks: One brood. Overwinters as partially grown larva. Adults prefer open wetlands and rarely leave transition zone around habitat. Named by Linnaeus from Philadelphia specimens.

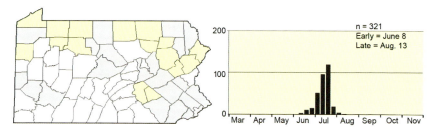

n = 321
Early = June 8
Late = Aug. 13

♂ ♀

average wing span

Distinguishing marks: Dorsal side hindwing with (1) terminal eyespots bearing central white dot (pupil). Ventral side light brown with (2) smooth dark postmedian line on both wings, compare with Eyed Brown (p. 164).

Typical behavior: Favors shady wet wooded areas and edges of shrub swamps. Found on sap, rotting fruit, dung, and moist soil.

Habitat: Shaded areas of forests and shrub swamps.

Larval hosts: Sedges, such as *Carex lacustris*, *C. amphibola*, and *C. grisea*.

Abundance: Uncommon, but not rare. Usually seen as singletons. S4

Remarks: One brood with partial second brood in southern counties. Overwinters as partially grown larva. Until the 1970s this species was considered a subspecies of the Eyed Brown (p. 164).

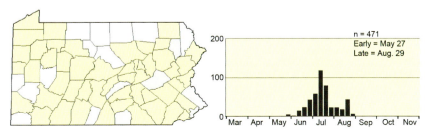

n = 471
Early = May 27
Late = Aug. 29

200

100

0
Mar Apr May Jun Jul Aug Sep Oct Nov

average wing span

Distinguishing marks: Dorsal side with (1) no eyespots. Ventral side with complete row of eyespots and (2) brown cell-end bar on both wings.

Typical behavior: Meanders in streamside vegetation. Flies low just over vegetation. Often darts back into woods.

Habitat: Shady moist woods, woodland edges, and streamsides with grasses.

Larval hosts: None reported. Uses several grasses in the South.

Abundance: Rare. S2

Remarks: Overwinters as partially grown larva. Found on six occasions between 1998 and 2015. May now be a breeding resident. Needs surveys for spring brood.

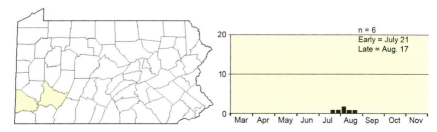

n = 6
Early = July 21
Late = Aug. 17

average wing span

Distinguishing marks: (1) Two yellow-ringed eyespots on both wings on both sides; female eyespots larger. Ventral hindwing intervening eyespots reduced to (2) silver scaling.

Typical behavior: Flight close to ground and bouncy. Visits sap, dung, and moist soil.

Habitat: Open woodlands, wood edges, fields, and damp areas along streams.

Larval hosts: Grasses such as orchard grass (*Dactylis glomerata*) and woodland muhly (*Muhlenbergia sylvatica*).

Abundance: Common throughout the state. S5

Remarks: One brood with bimodal emergence. Overwinters as partially grown larva. The early and late flights may represent two sibling species.

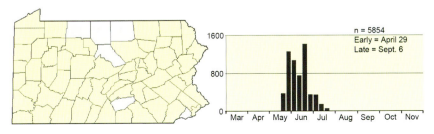

n = 5854
Early = April 29
Late = Sept. 6

♂   ♀

average wing span

Distinguishing marks: Dorsal side tan to rusty. Ventral side with single eyespot on forewing apex; heavy gray frosting, and jagged white line on hindwing.

Typical behavior: Flight close to ground and bouncy.

Habitat: Grassy open areas, including fields, meadows, and marshes.

Larval hosts: Grasses including fescue (*Festuca*) and little bluestem (*Schizachyrium scoparium*).

Abundance: Common. Found mostly in northern counties. S5

Remarks: Two broods. Overwinters as partially grown larva. Recently expanding southward from eastern Canada. Arrived in state in 1995 in Pike County. Now in twenty-six counties.

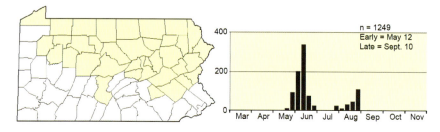

n = 1249
Early = May 12
Late = Sept. 10

ssp. *nephele*   ssp. *alope*

average wing span

Distinguishing marks: Dark brown with two large forewing eyespots. Eyespots situated in yellow patch (ssp. *alope*), in brown patch (ssp. *nephele*), or in one of many intermediate gradations. See Special Topic: Hybrid or Blend Zones (p. 158).

Typical behavior: Flight is low to ground and evasive when chased. Known to fall to ground and feign death.

Habitat: Wooded margins, open fields, streamsides, and thickets.

Larval hosts: Redtop (*Tridens flavus*). Also probably uses many other grasses.

Abundance: Common throughout the state. S5

Remarks: One prolonged brood. Males emerge before females. Overwinters as partially grown larva. Subspecies *alope* is more common in southern counties and *nephele* in northern counties. Forewing patch color highly variable.

n = 2261
Early = June 2
Late = Oct. 18

Τhe family Hesperiidae consists of small to medium-sized, stout-bodied butterflies commonly known as skippers. They get their common name from their quick erratic flight, seemingly skipping from place to place. They occur worldwide, with the greatest number of recognized species found in the tropics. About a third of North American butterflies belong to this family, and this is also true in Pennsylvania. There are four sub-families of Hesperiidae found in the state. These are the dicot skippers (Eudaminae), spread-winged skippers (Pyrginae), skipperlings (Heteropterinae), and grass skippers (Hesperiinae).

The adults are distinctly different from adults of other families. They have hairy, stocky bodies, strong thoracic muscles, large eyes, and antennae that terminate in hooks (see Special Topic: Apiculus, p. 231). Most of our skipper species feature drab colors varying from black to brown to tan; a few are more boldly marked with dashes of orange, yellow, or white. Skippers tend to rest with their wings partially or fully spread; the forewings and hindwings of one subfamily (grass skippers) are held in different planes (see Special Topic: "Jet Fighter" Wing Orientation, p. 199). Only a few species open the wings up completely.

Skippers are very active in daylight and actively take nectar from flowers. They are commonly found in gardens, meadows, forest edges, and just about everywhere flowers occur. Along with bees, they provide an important function as pollinators in ecosystems.

Viewed as a group, skipper larvae feed on a wide variety of hostplants. The larvae of dicot and spread-winged skippers feed on broad-leaved dicotyledonous plants in the bean, oak, willow, and mallow families. The larvae of the skipperlings and grass skippers feed on monocotyledonous plants in the grass and sedge families.

Females lay a hemispherical egg, usually singly, on leaves and stems of their hosts. Our skipper larvae are uniformly colored green or brown, virtually never brightly colored as are

The dicot skippers settle with their wings only partially open, as seen above in the photograph of the Silver-spotted Skipper.

the larvae of tropical species. All skipper larvae have a large head and a distinctive narrow ring ("collar") just behind the head. They typically feed at night and rest in leaf shelters during the day, hidden away from predators. Skippers typically hibernate in their leaf shelters over the winter. They ultimately pupate within their shelters or in nearby litter supported by a silk girdle.

Identifying skippers in the field can be frustrating, even for experienced naturalists. They are difficult to follow in flight and often stop for only short periods. Persistence and patience is needed to track them. Furthermore, there are several lookalike species within this group. Learning their subtle wing markings, habits, and hostplants ultimately helps to establish identifications. Mastering this group is a challenge, but it can also be a very satisfying endeavor.

Spread-winged skippers perch with their wings fully spread, as seen above with the male Wild Indigo Duskywing.

Dicot skippers gather nectar with their wings folded up, as seen above with the Hoary Edge (left) and the Silver-spotted Skipper (right).

The following seven species are dicot skippers (subfamily Eudaminae).

average wing span

Distinguishing marks: Dorsal side brown with (1) band of yellow rectangular spots on forewing. Ventral side with (2) prominent white patch on hindwing.

Typical behavior: Avid nectarer, fast flier with white patch on ventral side visible in flight.

Habitat: Meadows, fields, roadsides, gardens, forest edges, and trails.

Larval hosts: Generalist, using a wide variety of legumes, especially black locust (*Robinia pseudoacacia*), false indigo bush (*Amorpha fruticosa*), wild indigo (*Baptisia tinctoria*), hog peanut (*Amphicarpaea bracteata*), bush clovers (*Lespedeza*), and tick-trefoils (*Desmodium*).

Abundance: Common throughout the state. S5

Remarks: Multiple broods (usually three) with two large broods in summer. Pupa overwinters. Males perch on branches and low vegetation, darting out at intruders. Visits a large variety of nectar sources in summer.

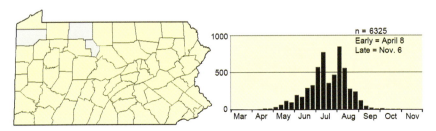

n = 6325
Early = April 8
Late = Nov. 6

♂   ♀

average wing span

Distinguishing marks: Dorsal side brown with white hyaline forewing spots; hindwing with long tails and blue-green iridescence. Ventral hindwing with dark brown stripes.

Typical behavior: Large robust skipper, most easily seen nectaring.

Habitat: Any open weedy area with nectar flowers; vacant lots, roadsides, etc.

Larval hosts: Variety of legumes, including tick-trefoils, American wisteria, and string beans.

Abundance: Uncommon migrant. SNA

Remarks: A southern skipper that rarely breeds in late summer in the state. Does not survive the winter in the northern states. Portions of one or both tails may be broken off.

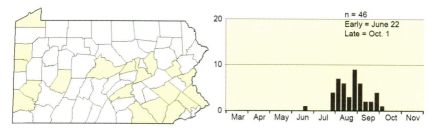

n = 46
Early = June 22
Late = Oct. 1

20

10

0

Mar Apr May Jun Jul Aug Sep Oct Nov

Golden-banded Skipper   *Autochton cellus*   (Boisduval and Le Conte, [18

average wing span

Distinguishing marks: Dorsal side brown with (1) bright gold band on forewing. Ventral side with gold band on forewing; hindwing with faint dark banding and no large white patch.

Typical behavior: Flight slow and near ground. Easily detected by bright gold band in flight.

Habitat: Moist woodland openings, trails, and ravines.

Larval hosts: Wild bean (*Phaseolus polystachios*).

Abundance: Rare. Possibly extirpated. SH*

Remarks: A southern skipper uncommon everywhere in its range. One brood, with partial second brood in some years. Pupa overwinters. Not reported since 1979. Needs surveys of potential habitat with newly clarified host (wild bean). Previously, host was incorrectly identified as hog peanut.

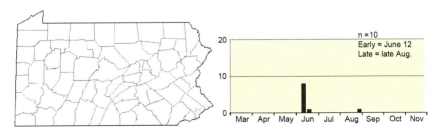

♂   ♀

average wing span

Distinguishing marks: Dorsal side brown with (1) offset yellow hyaline spots on forewing. Ventral side with (2) prominent white patch on outer hindwing. Fringe checkered black and white. Resembles Silver-spotted Skipper (p. 000), but white patch on hindwing reliably separates the two.

Typical behavior: Avid nectarer. Flies low to ground, not as fast as Silver-spotted Skipper.

Habitat: Open areas, such as forest edges and fields with nectar flowers.

Larval hosts: Tick-trefoils (*Desmodium canadense, D. paniculatum, D. rotundifolium*).

Abundance: Uncommon, local. S3

Remarks: One brood. Overwinters as full-grown larva. Look for this species in same areas as the more common Silver-spotted Skipper. Reduced numbers in recent years hint at statewide decline.

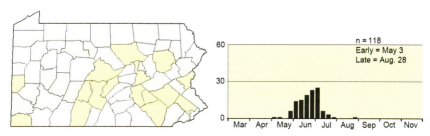

n = 118
Early = May 3
Late = Aug. 28

All three cloudywings occurring in the eastern United States have been recorded in Pennsylvania. Distinguishing the cloudywings can be difficult. They are very similar in phenotype and they vary a fair amount in their markings. A number of magnified views of important characteristics are presented in the Special Topic: Cloudywings (p. 180), following the individual species accounts.

average wing span

Distinguishing marks: Dorsal side forewing with (1) row of small white subapical spots on forewing, leading spot near costa faint or missing. Ventral side with (2) heavy gray frosting, especially on hindwing. Male has (3) costal fold. An additional set of Northern Cloudywing males appears on the following page showing variation in forewing spot pattern, and (4) unfurled costal fold. See Special Topic: Costal Fold for higher magnification of an unfurled fold (p. 181).

Typical behavior: Males perch on low vegetation, like shrubs and small trees, most of the day and aggressively defend territory.

Above male specimens show variations in forewing spot patterns.

Habitat: Woodland openings and edges, fields, meadows, and power line cuts with tick-trefoils and other legumes.

Larval hosts: Legumes, including tick-trefoils (*Desmodium*), bush clovers (*Lespedeza*), clovers (*Trifolium*), hog peanut (*Amphicarpaea bracteata*), black medick (*Medicago lupulina*), and alfalfa.

Abundance: Uncommon, but not rare. S3S4

Remarks: One brood. Full grown larva overwinters. Very pugnacious skipper within its territory, investigating and chasing invaders. Modest numbers in recent surveys hint at overall decline.

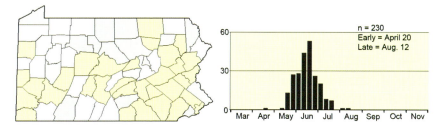

n = 230
Early = April 20
Late = Aug. 12

average wing span

Distinguishing marks: Dorsal side forewing with (1) row of white subapical spots, large and fixed in straight line; and (2) cluster of large white spots in median area. Ventral side mottled brown with light gray frosting on hindwing. Male lacks costal fold. White area often present on bend of antenna (see magnification above).

Typical behavior: Males perch and defend territory.

Habitat: Woodland openings and edges, fields, meadows, and power line cuts with tick-trefoils and other legumes.

Larval hosts: Legumes, including tick-trefoils (*Desmodium*) and bush clovers (*Lespedeza*).

Abundance: Uncommon, but not rare. Increasing in recent decades. S3S4*

Remarks: One brood, occasionally partial second brood. Fully grown larva overwinters. Starts flight slightly later than Northern Cloudywing (p. 176). Population moving northward in recent decades.

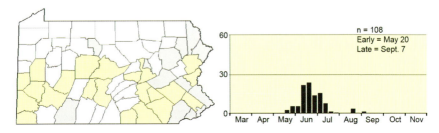

n = 108
Early = May 20
Late = Sept. 7

Confused Cloudywing   *Thorybes confusis*   E. Bell, 1923

average wing span

Distinguishing marks: Dorsal side forewing with (1) row of white subapical spots, broken with trailing spot offset outwardly; and cluster of intermediate-size white spots in median area. Ventral side mottled brown with light gray frosting on hindwing. (2) Palps very lightly colored and white eye ring present. Male lacks costal fold.

Typical behavior: Males perch and defend territory.

Habitat: Fields, meadows, and open areas near woods.

Larval hosts: Legumes, including various tick-trefoils (*Desmodium*) in the south.

Abundance: Rare stray. Recorded infrequently in state. SNA

Remarks: A southern skipper that does not breed in the state. Difficult to distinguish from other cloudywings due to phenotypic variability and overlap. Genitalic dissection by an expert is often needed for conclusive identification.

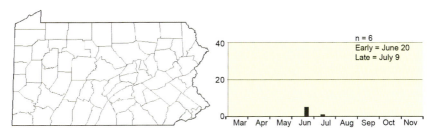

n = 6
Early = June 20
Late = July 9

Northern Cloudywing   Southern Cloudywing   Confused Cloudywing

Male with costal fold

Male with no costal fold

Male with no costal fold

*Small spots, leading edge
faint to missing*

*Large spots,
nearly fused*

*Medium spots,
trailing edge offset*

*Under, side, and top of head:
dark brown brown palps
and no white eye ring*

*Under, side, and top of head:
grayish palps,
thin white eye ring*

*Under, side, and top of head:
white palps with
white eye ring*

The term *costa* denotes the leading edge of a wing. A fold in the forewing costa occurs in some male skippers. This fold contains scent scales used to attract females; it is one means of distinguishing males from females. Furthermore, since it does not occur in all skippers, its presence or absence can be used to help differentiate closely related species. For example, the male Northern Cloudywing is the only cloudywing occurring in Pennsylvania that has a costal fold. The photos below show the costal fold of the Northern Cloudywing in a folded (top) and unfolded (bottom) state.

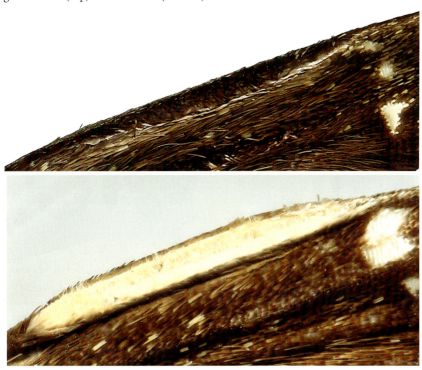

Duskywings   *Erynnis*

The following ten skipper species are known as duskywings (genus *Erynnis*). Like cloudywings, duskywings are similar to one another in appearance and their markings may vary, making their identification difficult. Duskywings belong to the subfamily Pyrginae, commonly known as spread-winged skippers.

When trying to identify duskywing species, besides carefully examining their markings, it is very useful to make note of their flight periods and hostplants. Several species are seen only in the spring and most are not found far from their hostplants. In some cases, consultation with experts and genitalic dissection may be needed for a positive identification.

A comparative chart of flight periods and phenotypic markings is presented in Special Topic: Duskywing Identification (p. 192), following the individual species accounts.

The following fourteen species are spread-winged skippers (subfamily *Pyrginae*).

♂   ♀

average wing span

Distinguishing marks: Smallest duskywing. (1) Labial palps elongated. Dorsal forewing lacks subapical white spots. Female forewing paler with more distinct bands. Male with costal fold (as in all duskywings).

Typical behavior: Males perch on low vegetation in sunny spots awaiting females. Males mineralize on dirt roads in woodlands with other duskywings.

Habitat: Open woodlands, woodland edges, trails, and roadsides.

Larval hosts: Gray birch (*Betula populifera*) and quaking aspen (*Populus tremuloides*). Rarely, uses black locust (*Robinia pseudoacacia*) in xeric habitats.

Abundance: Common throughout the state. S4

Remarks: One brood. Overwinters as fully grown larva. Only Dreamy and Sleepy Duskywings (p. 183) lack subapical white spots on forewing. Smaller size and elongated labial palps help distinguish Dreamy from Sleepy Duskywing.

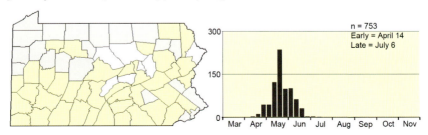

n = 753
Early = April 14
Late = July 6

Sleepy Duskywing *Erynnis brizo brizo* (Boisduval and Le Conte, [1837])

♂ ♀

average wing span

Distinguishing marks: Larger than Dreamy Duskywing (p. 182) and labial palps shorter. Dorsal forewing lacks subapical white spots. Female forewing paler with more distinct bands.

Typical behavior: Males perch on low tree limbs or bare twigs awaiting females. Males mineralize on dirt roads in woodlands with other duskywings.

Habitat: Open oak forest, forest edges, trails, and roadsides.

Larval hosts: Oaks (*Quercus alba*, *Q. ilicifolia*, *Q. prinus*). Also observed ovipositing on sprouts of American chestnut (*Castanea dentata*).

Abundance: Common throughout the state. S4

Remarks: One brood. Overwinters as fully grown larva. Peak emergence is slightly earlier than Dreamy Duskywing. Larger size and shorter labial palps help distinguish Sleepy from Dreamy Duskywing.

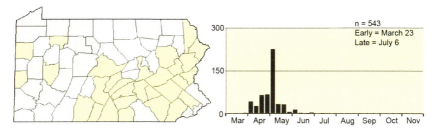

n = 543
Early = March 23
Late = July 6

♂   ♀

average wing span

Distinguishing marks: Large duskywing. Dorsal forewing with (1) a row of three or four hyaline white subapical spots; below them one to three postmedian white spots and a discal cell end spot, larger in the female. Ventral side with (2) two small white spots near leading edge of hindwing, often reduced in male.

Typical behavior: Males perch on low tree limbs and bare twigs awaiting females. Males mineralize at damp soil with other duskywings.

Habitat: Open oak forests, forest edges, trails, and roadsides.

Larval hosts: Oaks (*Quercus alba, Q. rubra*). Likely uses other oak species.

Abundance: Common throughout the state. S5

Remarks: One brood. Overwinters as fully grown larva. Typically the most common woodland duskywing in spring. To distinguish Juvenal's from the similar Horace's Duskywing (p. 185), examine the underside of the hindwing for two white spots.

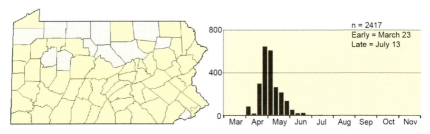

n = 2417
Early = March 23
Late = July 13

♂ ♀

average wing span

Distinguishing marks: Similar size as Juvenal's Duskywing (p. 184). Dorsal forewing with three or four hyaline white subapical spots; below them one to three postmedian white spots and a discal cell end spot. Ventral hindwing lacking white spots.

Typical behavior: Males perch on low tree limbs and bare twigs awaiting females. Males mineralize at damp soil with other duskywings.

Habitat: Open oak forests, forest edges, trails, and roadsides.

Larval hosts: Oaks (*Quercus*).

Abundance: Common. Missing from north-central Pennsylvania. S4

Remarks: Two broods and sometimes a partial third brood. Overwinters as fully grown larva. Found in association with Juvenal's Duskywing, but less common. To distinguish Horace's from the similar Juvenal's Duskywing, examine the ventral hindwing for absence of two white spots.

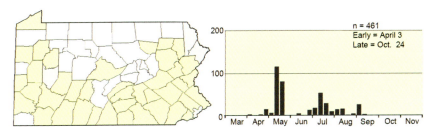

n = 461
Early = April 3
Late = Oct. 24

average wing span

Distinguishing marks: Dorsal side with heavily mottled wings; forewing white spots somewhat obscured by mottling. Ventral side with alternating light and dark brown dots.

Typical behavior: Males perch on low tree limbs and bare twigs awaiting females. Males mineralize at damp soil with other duskywings.

Habitat: Open areas such as open woodlands, meadows, and serpentine barrens where New Jersey tea grows.

Larval hosts: New Jersey tea (*Ceanothus americanus*).

Abundance: Rare. S1*

Remarks: Two broods. Overwinters as fully grown larva. Rapidly declining in state along with the disappearance of *Ceanothus*. Last recorded in 2005.

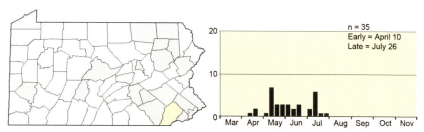

n = 35
Early = April 10
Late = July 26

♂   ♀

average wing span

Distinguishing marks: Forewing narrower and more pointed than that of other duskywings. Dorsal forewing with (1) subapical spots that are thin and oblong; and discal cell end white spot faint to absent. Resembles Wild Indigo Duskywing (p. 190), but has narrower, pointed forewings.

Typical behavior: Nectars frequently.

Habitat: Open areas, fields, wood edges, and roadsides.

Larval hosts: Reported on black locust (*Robinia pseudoacacia*), bush clovers (*Lespedeza*), and wild indigos (*Baptisia*) in the South.

Abundance: Infrequent migrant. SNA

Remarks: Southern species that seldom appears in late summer. Last recorded in state in 1982. Very similar in appearance to the ubiquitous Wild Indigo Duskywing. Perhaps occurs more frequently than recorded.

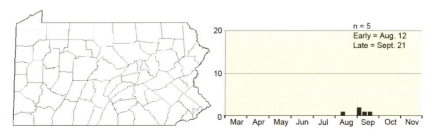

n = 5
Early = Aug. 12
Late = Sept. 21

♂   ♀

average wing span

Distinguishing marks: Forewing narrower and more pointed than other duskywings with exception of Zarucco Duskywing (p. 187). Hindwing with prominent white fringe. The flashy white hindwing fringe separates it from other duskywings in our area.

Typical behavior: Nectars frequently.

Habitat: Open areas, fields, wood edges, and roadsides.

Larval hosts: Uses many different legumes in the Southwest.

Abundance: Rare migrant. SNA

Remarks: Resident in a range from the American Southwest to South America. Known for occasional wide dispersals. Last recorded in state in 2003.

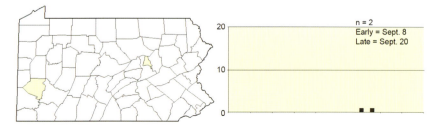

n = 2
Early = Sept. 8
Late = Sept. 20

average wing span

Distinguishing marks: Dorsal forewing with (1) subapical spots in zigzag row. Males with moderate amount of fine white hairlike scales on dorsal forewing. See Persius Complex in Special Topic: Duskywing Identification (p. 192).

Typical behavior: Males perch on low vegetation in vicinity of wild columbine awaiting females.

Habitat: Woodland trails, woodland edges, hillsides, and limestone outcroppings where wild columbine grows.

Larval hosts: Wild columbine (*Aquilegia canadensis*).

Abundance: Rare. S1S2*

Remarks: Two broods with occasional third brood. Overwinters as fully grown larva. Adults do not stray far from larval host plant. Has been known to wander into yards and oviposit on garden columbines. Last recorded in state in 2008.

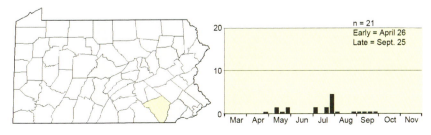

n = 21
Early = April 26
Late = Sept. 25

Wild Indigo Duskywing   *Erynnis baptisiae*   (W. Forbes, 1936)

♂   ♀

average wing span

Distinguishing marks: Larger than Columbine and Persius Duskywings (pp. 189, 191). (1) Dorsal forewing subapical spots in zigzag row. Males with scanty amount of fine white hairlike scales. See Persius Complex in Special Topic: Duskywing Identification (p. 192).

Typical behavior: Males perch on low vegetation in vicinity of crown vetch awaiting females. Males mineralize at damp soil with other duskywings.

Habitat: Open areas, fields, meadows, roadsides, right-of-ways, railroad tracks, interstate high-way shoulders, parks, and virtually anywhere crown vetch grows.

Larval hosts: Crown vetch (*Securigera* [=*Coronilla*] *varia*), wild indigo (*Baptisia tinctoria, B. australis*), white sweet clover (*Melilotus albus*), and golden clover (*Trifolium aureum*).

Abundance: Very common throughout the state. S5

Remarks: Three broods. Partial fourth brood in some years. Overwinters as fully grown larva. Prior to the 1960s this species was confined to barrens with wild indigo. Now it is ubiquitous in all habitats where crown vetch has spread.

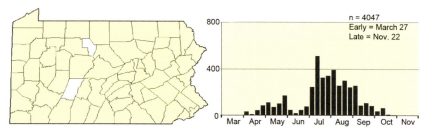

n = 4047
Early = March 27
Late = Nov. 22

♂   ♀

average wing span

Distinguishing marks: Dorsal forewing with (1) white apical spots aligned in nearly straight line. Males with numerous projecting white hairlike scales on dorsal forewing. See Persius Complex in Special Topic: Duskywing Identification (p. 192).

Typical behavior: Males perch on low vegetation in open dry areas with wild indigo.

Habitat: Scrub oak barrens and woodland openings where larval host grows.

Larval hosts: Wild indigo (*Baptisia tinctoria*). Wild lupine (*Lupinus perennis*) was also a host earlier in the twentieth century.

Abundance: Rare. S1S2*

Remarks: One brood. Overwinters as fully grown larva. Presently found only in scrub oak barrens associated with wild indigo. Last recorded in 2000.

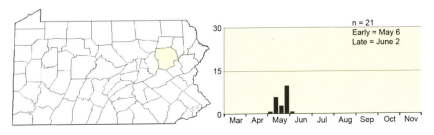

n = 21
Early = May 6
Late = June 2

The ten species of duskywings (genus *Erynnis*) recorded in Pennsylvania, as a group, are very difficult to identify to species level. To help in their identification, the following pages present a direct comparison of the size, markings, and flight periods of resident species. (The rare migrants Funereal and Zarucco Duskywings are not included.)

Three closely related species make up the Persius Complex. This group is distinguished by fine white hairlike scales on the dorsal forewing. Note that there are two kinds of small white

Dreamy Duskywing

Sleepy Duskywing

Juvenal's Duskywing

Horace's Duskywing

Mottled Duskywing

Apr May Jun Jul Aug Sep Oct

forewing scales (see bottom photos). One is a rectangular scale with a cut-off distal edge seen in many skippers; the other is a long, thin, tapering hairlike scale found in the Persius Complex. The Persius Duskywing (left) has the largest amount of hairlike scales, while the Columbine Duskywing (middle) and Wild Indigo Duskywing (right) have decreasing amounts.

Apr May Jun Jul Aug Sep Oct

Persius Duskywing

Columbine Duskywing

Wild Indigo Duskywing

♂   ♀

average wing span

Distinguishing marks: Small dark skipper with (1) distinctly scalloped wing margins. Male with (2) costal fold (unfolded in above photo).

Typical behavior: Perches low to the ground with wings spread. Often overlooked.

Habitat: Waste areas, old fields, roadsides, right of ways.

Larval hosts: Lamb's quarters (*Chenopodium album*).

Abundance: Uncommon resident in few southern counties. S3S4

Remarks: Two broods. Overwinters as fully grown larva in a nest of dead leaves. Species is migratory and some individuals enter state from the south.

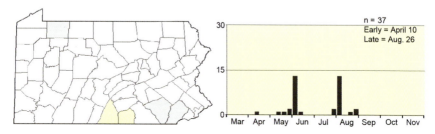

n = 37
Early = April 10
Late = Aug. 26

♂   ♀

average wing span

Distinguishing marks: Black-and-white checkered skipper. Slightly smaller and darker than Common Checkered-Skipper (p. 196). Lacks blue-gray dorsal sheen.

Typical behavior: Flight is close to ground in vicinity of hostplant. Males continually patrol for females.

Habitat: South- to west-facing rock outcrops in Appalachian foothills, particularly shale barrens.

Larval hosts: Dwarf cinquefoil (*Potentilla canadensis*).

Abundance: Rare. Very localized colonies. S1*

Remarks: One brood (spring). Overwinters as pupa. Declining in Pennsylvania in recent years. Previously occurred in open grassy limestone-based hillsides in southeast portion of state. Now restricted to Ridge and Valley region. A skipper specialist that utilizes only dwarf cinquefoil as its larval host.

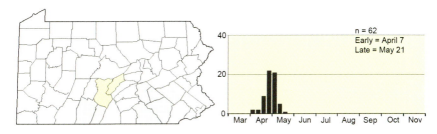

n = 62
Early = April 7
Late = May 21

♂ ♀

average wing span

Distinguishing marks: The only common black-and-white checkered-skipper in the state. Male has a blue-gray dorsal sheen when fresh.

Typical behavior: Males patrol open brushy areas with mallows, searching for females.

Habitat: Pastures, weedy fields, meadows, roadsides, and gardens.

Larval hosts: Common mallow (*Malva neglecta*), round-leaved mallow (*M. rotundifolia*), marsh mallow (*Althaea officinalis*), and hollyhock (*Alcea*).

Abundance: Common migrant. Year-round resident in South and Southwest. S4S5B

Remarks: One to three broods depending upon when temporary populations are established. The greatest abundance is reached from late July to October. Does not survive winters in Pennsylvania. The nearly identical White Checkered Skipper (*P. albescens*) of the Southwest has recently invaded the Southeast. Confirmation of a stray into Pennyslvania would require an expert.

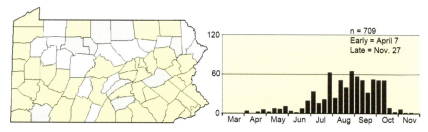

n = 709
Early = April 7
Late = Nov. 27

♂   ♀

average wing span

Distinguishing marks: Common small, dark skipper. Wing margins smooth, not scalloped. Forewing with white dots. Hindwing unmarked. Both sexes with white palpi. Male with costal fold.

Typical behavior: Flight is low and erratic, usually in wastelands such as vacant lots and land-fills.

Habitat: Weedy fields, vacant lots, disturbed areas, streamsides, and roadsides.

Larval hosts: Lamb's quarters (*Chenopodium album*), white pigweed (*Amaranthus albus*), and feather cockscomb (*Celosia argentea*).

Abundance: Common resident in southern counties. S4S5

Remarks: Three broods. Overwinters as fully grown larva. This small skipper has benefited from human-altered landscapes. Southern migrants often enhance summer numbers.

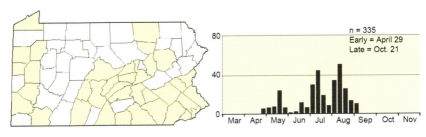

n = 335
Early = April 29
Late = Oct. 21

Arctic Skipper *Carterocephalus palaemon mandan* (W. H. Edwards, 1863)

The following species is the only skipperling (subfamily Heteroptinae) in the state.

♂   ♀

average wing span

Distinguishing marks: Small skipper. Dorsal side dark brown with orange spots. Ventral side orange with white spots on hindwing.

Typical behavior: Males perch in sunlit low vegetation in vicinity of northern wetlands.

Habitat: Edges of wetlands, marshes, and grassy streamsides.

Larval hosts: Blue joint grass (*Calamagrostis canadensis*).

Abundance: Uncommon. Found in local colonies in northern counties. S3*

Remarks: One brood. Overwinters as fully grown larva. This skipper ranges from subarctic North America to Eurasia. The southern extent of its range in eastern North America occurs in northern Pennsylvania.

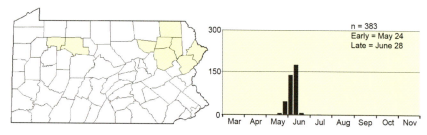

n = 383
Early = May 24
Late = June 28

The grass skippers make up almost two-thirds of skippers found in Pennsylvania. They are small, generally orange-brown or brown, bordered in black, and decorated with yellow or white spots. As a group they lack highly distinctive markings and there is often considerable variation within individual species. For these reasons their identification is sometimes difficult. In the following pages, more than one male and female have been pictured for some species. Also, special topics have been interspersed between species accounts to highlight important characteristics. Even with these aids, some species continue to confuse experienced experts.

Eggs are generally laid singly and their larvae feed on grasses or sedges. In many species larvae feed at night and rest during the day in shelters made of bent and folded grass leaves. They often overwinter and pupate the following spring in these shelters.

Males lack the costal fold that is commonly found in spread-winged skippers. Males of a majority of species possess a stigma, a group of androconial or scent scales on the dorsal side of the forewing. Stigmata are important in terms of identification (see Special Topic: Stigmata of Skippers, p. 226). In the past these stigma-bearing skippers were known as branded skippers.

When resting, their wings are folded up. When basking in the sun or nectaring, they adapt a characteristic "jet fighter" pose, with hindwings flat and forewings at a forty-five-degree angle (see Special Topic: "Jet Fighter" Wing Orientation, p. 199).

Special Topic   "Jet Fighter" Wing Orientation

Shown to the left below is a head-on view of the typical posture of a grass skipper, known as the "jet fighter" orientation. Note the forewings are at a forty-five-degree angle to the hindwings, which are spread out flat. Grass skippers commonly adopt this position when nectaring or basking.

When photographing or observing a skipper in this position, one should try and align oneself with one of the forewings, thereby looking directly at the opposite forewing and downward at the same side hindwing. This approach maximizes the amount of observable information on the dorsal side, which ultimately aids in the identification of the skipper. An example of this method is shown on the right below, featuring a male Delaware Skipper.

The following thirty-five species are grass skippers (subfamily Hesperiinae).

average wing span

Distinguishing marks: Dorsal side plain dark brown. Rare individuals with pale spots in mid-forewing. Ventral side brown with veins highlighted by yellowish scales (see magnified views).

Typical behavior: Males perch on low vegetation awaiting females.

Habitat: Grassy fields, meadows, power line cuts, streamsides.

Larval hosts: Little bluestem (*Schizachyrium scoparium*) and broomsedge bluestem (*Andropogon virginicus*).

Abundance: Uncommon. More common in southern counties. Prone to population fluctuations. S3S4*

Remarks: Two broods. Overwinters as fully grown larva. When observing a small drab skipper with virtually no markings, consider the Swarthy Skipper or male Dun Skipper (p. 237).

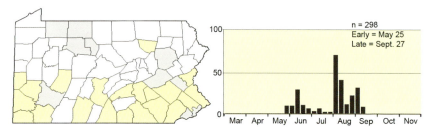

n = 298
Early = May 25
Late = Sept. 27

average wing span

Distinguishing marks: Dorsal side forewing with (1) small white spots in subapical region, bottom one often shifted toward apex; male with (2) black stigma; and female with (3) prominent central white spots. Ventral side with (4) violet-gray scaling toward outer margin of both wings.

Typical behavior: Males perch on or near ground awaiting females.

Habitat: Open grassy areas, fields, forest edges, and streamsides.

Larval hosts: Not reported to breed in state. Resident of southern United States, where it uses a variety of grasses.

Abundance: Uncommon migrant. SNA

Remarks: Appears periodically. When present, it typically arrives in August and September. Other skippers with violet-gray scaling on ventral hindwings include the Dusted, female Zabulon, and Common Roadside (pp. 240, 229, 242). (The Dusted Skipper flies in spring, the female Zabulon has white leading edge of hindwing, and the Common Roadside-Skipper is smaller). See Special Topic: Fall Migrants (p. 211) for added examples.

n = 35
Early = May 5
Late = Oct. 25

♂   ♀

average wing span

Distinguishing marks: Our smallest skipper. Dorsal forewing with variable orange; hindwing orange with wide black border. Ventral hindwing solid orange; forewing black with orange costa and apex. Antenna short.

Typical behavior: Slow, weak flight, low to ground among grasses. Unlike most grass skippers, Least Skipper males are patrollers. They weave and meander through grasses seeking females.

Habitat: Wet open areas, marshes, streamsides, wet meadows, ditches.

Larval hosts: Reed canary grass (*Phalaris arundinacea*) and rice cutgrass (*Leersia oryzoides*). Probably feeds on many grass species.

Abundance: Common throughout the state. S5

Remarks: Three broods. Overwinters as partially grown larva.

n = 2610
Early = April 24
Late = Nov. 8

♂   ♀

average wing span

Distinguishing marks: Dorsal forewing brassy orange with black veins and border; hindwing with black toning, but without large border as in Least Skipper. Ventral side dull yellowish orange, not bright yellow or orange as in Delaware Skipper (p. 225). Fringes tan. Antenna short.

Typical behavior: Avid nectarer. Flight is low to the ground. Frequently perches on grass blades.

Habitat: Open areas, pastures, meadows, fields, roadsides.

Larval hosts: Timothy grass (*Phleum pratense*).

Abundance: Common throughout the state. Can be abundant at peak flight. S5

Remarks: One brood. Overwinters as egg. An introduced Eurasian species, first reported in Pennsylvania in 1953. Seen from below it resembles the Delaware Skipper. The latter is larger with more intense ventral color.

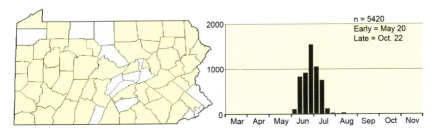

n = 5420
Early = May 20
Late = Oct. 22

♂ ♀

average wing span

Distinguishing marks: Dorsal side of male orange with (1) jagged black border. Female predominantly dark brown with fewer orange markings. Ventral hindwing light orange (male) to tawny (female) with black dots of varying size. Short, curved antennae. See next page and Special Topic: Fall Migrant Skippers (p. 211) for other examples of Fiery Skippers.

Typical behavior: Males perch on low vegetation awaiting females.

Habitat: Open areas, fields, meadows, gardens, lawns, and parks.

Larval hosts: Crabgrass (*Digitaria*). Probably feeds on many grass species.

Abundance: Common migrant. S4B

Remarks: Southern migrant. Typically arrives in late summer and fall. One to two broods, once established. Does not survive winter in Pennsylvania. The ventral side of the Whirlabout (p. 220) is similar, but has a slightly different spot pattern.

n = 864
Early = May 5
Late = Nov. 7

Above examples depict the phenotypic variation occurring in female Fiery Skippers. Below are dorsal and ventral views of Fiery Skippers seen in the wild; top row males, bottom row females.

average wing span

Distinguishing marks: Dorsal forewing of male with (1) large square stigma. Female forewing with (2) two large hyaline spots. Ventral hindwing of male light yellowish brown with faint markings; female with (3) pronounced *V*-shaped row of pale spots (chevron). See next page and Special Topic: Fall Migrant Skippers (p. 211) for additional examples of Sachems.

Typical behavior: Males perch on low vegetation awaiting females.

Habitat: Open areas, fields, meadows, gardens, lawns, and parks.

Larval hosts: Orchard grass (*Dactylis glomerata*), Bermuda grass (*Cynodon dactylon*), goose grass (*Eleusine indica*), and crabgrass (*Digitaria*).

Abundance: Very common migrant. S5B

Remarks: Southern migrant. Typically arrives in late summer and fall. One to three broods, once established. Does not survive winter in Pennsylvania. When both sexes nectar at the same flower, they are often confused as being two different species.

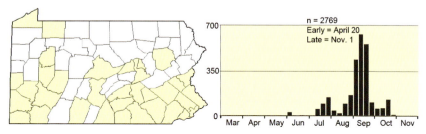

n = 2769
Early = April 20
Late = Nov. 1

Above examples depict the phenotypic variation occurring in male and female Sachems. Below are dorsal and ventral views of Sachems seen in the wild; top left is male, remaining are females. Note the difference between fresh and worn individuals (bottom two).

average wing span

Distinguishing marks: Dorsal forewing of male with (1) thin black stigma with central white line, common to genus *Hesperia*. Female forewing darker with diagonal row of distinct yellow spots. Ventral side reddish brown with (2) chevron pattern of yellow to buff-colored spots on hindwing.

Typical behavior: Males perch waiting for females. Flight is very fast when darting out to investigate subjects or when flying from flower to flower.

Habitat: Open areas, old fields, roadsides, barrens.

Larval hosts: Switchgrass (*Panicum virgatum*), lovegrass (*Eragrostis*), poverty oat grass (*Danthonia*), and bluestem (*Schizachyrium*).

Abundance: Uncommon. S3S4*

Remarks: One brood in late summer. Overwinters as young larva (first instar). The ventral sides of a few female skippers are similar. Helpful hints to distinguish them from Leonard's Skipper are: Peck's Skipper (p. 214) is smaller; Cobweb Skipper, Indian Skipper, and Long Dash (pp. 209, 212, 219) have finished flight. The female Sachem (p. 206) has two large hyaline spots in center of forewing (best seen with wings open).

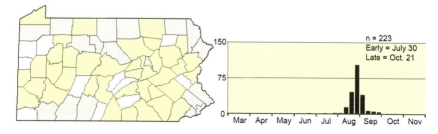

n = 223
Early = July 30
Late = Oct. 21

♂ ♀

average wing span

Distinguishing marks: Dorsal side with scant orange; male with thin black stigma with central white line; female darker with larger forewing white spots. Ventral side brown with cream-colored chevron pattern on hindwing; white veins produce a "cobweb" appearance.

Typical behavior: Males perch on low vegetation awaiting females.

Habitat: Shale barrens, old fields, open scrub areas.

Larval hosts: Little bluestem (*Schizachyrium scoparium*).

Abundance: Uncommon. Very local and unpredictable. S2S3*

Remarks: One brood in early spring. Overwinters as fully grown larva. Populations noncontinuous (disjunct) throughout range.

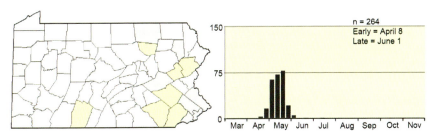

n = 264
Early = April 8
Late = June 1

average wing span

Distinguishing marks: Dorsal forewing of male orange and brown; female darker with pale spots. Male with black stigma with central white line. Ventral hindwing greenish brown with curved row of white spots ("dotted").

Typical behavior: Males perch on low vegetation awaiting females. Fond of knapweed as nectar source.

Habitat: Sunny, grassy, unmowed fields; dry meadows; sandy pine forests.

Larval hosts: Does not breed in Pennsylvania. Uses switchgrass (*Panicum virgatum*) elsewhere.

Abundance: Rare stray. SNA

Remarks: A stray individual was recorded once in southeastern Pennsylvania in 1962. Local colonies persist in southern New Jersey.

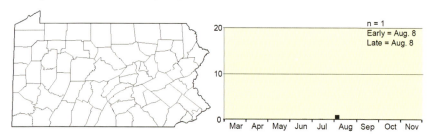

n = 1
Early = Aug. 8
Late = Aug. 8

Several skippers migrate northward into the state in late summer and fall. The Sachem and Fiery Skipper are common annual visitors. Less common migrants include the Ocola Skipper, Long-Tailed Skipper, and Clouded Skipper. Even rarely seen skippers like the Zarucco Duskywing, Funereal Duskywing, Whirlabout, Southern Broken-Dash, Twin-spot, and Brazilian Skippers may stray into the state.

Some migrants may be difficult to distinguish from resident species. Shown below, in the top row, are the common migrant Fiery Skipper (left) and the much less common Whirlabout (right). Because of their similarity, Whirlabouts may be overlooked. Butterfly enthusiasts in the southeastern portion of the state should be on the lookout for this species in the fall. The middle row shows two views of the less common migrant Clouded Skipper. Its ventral side may be confused with female Zabulon Skipper and dorsal side with one of the female Witches (see Special Topic: Three Witches, p. 239). The bottom row shows two views of the very distinctive elongated forewing of the Ocola Skipper.

♂   ♀

average wing span

Distinguishing marks: Dorsal side orange and black with inwardly toothed borders. Male with black stigma with white central line and surrounded by orange. Ventral side dull orange with chevron pattern of square yellow spots on hindwing. See images on the following page for examples of pattern variation on ventral hindwing.

Typical behavior: Males perch on low vegetation awaiting females. Males dart out to inspect insects flying by.

Habitat: Woodland openings and edges, brushy fields.

Larval hosts: Poverty oat grass (*Danthonia spicata*). Likely feeds on other grasses. Reported on little bluestem (*Schizachyrium scoparium*) elsewhere.

Abundance: Uncommon. In some localities it is readily found. S3S4*

Remarks: One brood. Overwinters as partially grown larva. See Remarks under Leonard's Skipper for discussion of species with similar ventral hindwings.

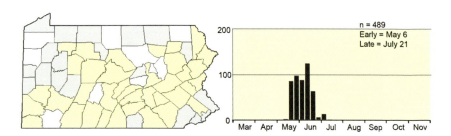

n = 489
Early = May 6
Late = July 21

♂　　　　♀

The pair above depict the less-conspicuous ventral markings found in Indian Skippers. With reduced orange and enhanced dark markings, the female may resemble other grass skippers, such as Peck's Skipper and Long Dash (pp. 214, 219). When the venter is awash with yellow overscaling, the chevron pattern is difficult to see.

Below are examples of typical Indian Skippers in the field.

average wing span

Distinguishing marks: Dorsal forewing of male with (1) orange above and extensive black patchlike area below broad stigma. Female forewing with prominent dark line through cell. Hindwing with (2) small orange area divided by black veins. Ventral hindwing with (3) two rows of rectangular yellow spots; middle spot of outer row elongated beyond other spots; the two rows occasionally touch.

Typical behavior: Avid nectarer. Males perch on low vegetation awaiting females.

Habitat: Woodland edges, old fields, roadsides, vacant lots, parks, gardens.

Larval hosts: Orchard grass (*Dactylis glomerata*). Likely feeds on other grasses.

Abundance: Common throughout the state. S5

Remarks: Two to three broods. Partial fourth brood in some years. Overwinters as fully grown larva or pupa. Ventral hindwing of Long Dash (p. 219) is similar to Peck's Skipper; however, middle spot of outer row is not as elongated as Peck's and there is not a complete inner row.

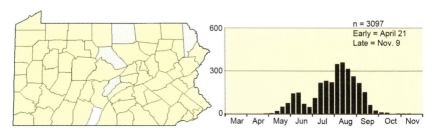

♂ ♀

The examples above illustrate the variation in degree of dorsal melanization and the size of the yellow cells on ventral hindwing.

The bottom examples show variable dorsal patterns of females (top) and males (below) in the wild.

Female Tawny-edged and Crossline Skippers are sometimes grouped with females of other grass skippers termed the "Three Witches." Forewing postmedian spot patterns are helpful in distinguishing these females (see Special Topic: Three Witches, p. 239). Female Tawny-edged and Crossline Skippers are also distinguished by the degree of orange on the leading edge of the dorsal forewing. Males have different styles of stigmata, which are shown in detail in Special Topic: Stigmata of Skippers (p. 226). Below are additional examples showing the differences between these two species, as well as illustrating the range of coloring within each species when compared to the previous specimens.

Tawny-edged Skipper

Crossline Skipper

average wing span

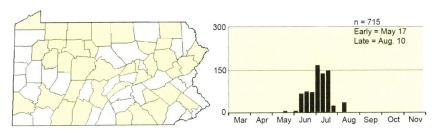

Distinguishing marks: Dorsal forewing of male with (1) black stigma nearly surrounded by orange and touching black subapical spot. Female forewing with multiple polygonal orange spots. Ventral hindwing with curved row of orange rectangles and a basal spot.

Typical behavior: Avid nectarer. Males perch on low vegetation awaiting females.

Habitat: Woodland edges; streamsides; low, damp, grassy areas.

Larval hosts: Bluegrass (*Poa*). Likely feeds on other grasses.

Abundance: Uncommon. More frequently met with in northern counties. S3S4*

Remarks: One brood. Overwinters as partially grown larva. In the male, the "long dash" consists of black stigma and black subapical spot extending across entire forewing to the dark border. Some similarity to Indian Skipper (p. 212), above and below.

n = 715
Early = May 17
Late = Aug. 10

average wing span

Distinguishing marks: Dorsal forewing of male with (1) stigma surrounded by orange and faint black markings; and (2) thin black border. Female forewing with multiple polygonal cream-colored spots. Ventral side yellow (male) or grayish yellow (female) with submarginal and basal brown spots.

Typical behavior: Adults land and take off in a rapid circular flight.

Habitat: Open areas, old fields, pastures, roadsides.

Larval hosts: Does not breed in Pennsylvania. Uses many grasses in the South.

Abundance: Rare migrant. Usually follows the Coastal Plain in the fall. SNA

Remarks: A southern species. Periodically appears in southeastern Pennsylvania; last taken in 1965. Resembles the common fall migrant Fiery Skipper (p. 204); the ventral hindwings of the Whirlabout have fewer, but more-distinct brown spots. It is worth a second look at Fiery Skippers during their fall migration to detect a possible rare Whirlabout. See further comparison in Special Topic: Fall Migrants (p. 211).

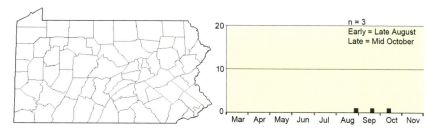

n = 3
Early = Late August
Late = Mid October

average wing span

Distinguishing marks: Dorsal forewings of both sexes with diagonal row of pearly white translucent spots ("glassy wing"); central spot next to male stigma (1) trapezoidal, and in female (2) squarish. Ventral side dark brown with purplish sheen when fresh; hindwing with arc of faint white spots. Antenna with (3) patch of white scales immediately below club.

Typical behavior: Avid nectarer. Males perch on low vegetation awaiting females.

Habitat: Woodland edges, old fields, roadsides.

Larval hosts: Redtop (*Tridens flavus*). Likely feeds on other grasses.

Abundance: Common throughout the state. S5

Remarks: One brood. Overwinters as early-stage larva. The female Little Glassywing is similar to other small dark female skippers that fly at the same time (Northern Broken-Dash, Dun Skipper, pp. 222, 237), making them difficult to distinguish. Collectively, they are often referred to as the "Three Witches." See Special Topic: Three Witches (p. 239).

n = 2035
Early = May 23
Late = Sept. 13

average wing span

Distinguishing marks: Dorsal forewing of male with slight orange on leading edge; a multiply divided stigma ("broken dash"); and (1) prominent yellow-orange spot at end of stigma. Female forewing with diagonal row of light yellow spots with (2) two central ones rectangular. Ventral side dark brown with purplish sheen when fresh. Hindwing with arc of indistinct yellow-white spots; middle spot sometimes projects inwardly, creating a weakly shaped 3; occasional faint spot basally.

Typical behavior: Avid nectarer. Males perch on low vegetation awaiting females.

Habitat: Open grassy areas, old fields, roadsides, parks.

Larval hosts: Deer tongue (*Dichanthelium clandestinum*) and western panic grass (*D. acuminatum*).

Abundance: Common throughout the state. S5

Remarks: One brood with partial second on Atlantic coastal plain. Overwinters as partially grown larva. Worn specimens may appear orangish on venter, but never as deep orange as Southern Broken-Dash (p. 223). The female Northern Broken-Dash is similar to other small dark skippers (Dun Skipper, Little Glassywing, pp. 237, 221) that fly at the same time and are difficult to distinguish. Collectively, they are frequently referred to as the "Three Witches." See Special Topic: Three Witches (p. 239).

n = 1448
Early = May 30
Late = Sept. 13

♂   ♀

average wing span

Distinguishing Marks: Similar to Northern Broken-Dash (p. 222), except more reddish brown. Male dorsal forewing with multiply divided stigma ("broken dash") and (1) prominent orange spot at end of stigma. Female forewing with diagonal row of yellow-orange spots with (2) two central ones rectangular. Ventral hindwing with (3) arc of indistinct yellow-orange spots; middle spot often projects inwardly, creating a weakly shaped 3.

Typical behavior: Avid nectarer.

Habitat: Open grassy areas, woodland edges, streamsides.

Larval hosts: Does not breed in Pennsylvania. Uses many grasses in the South.

Abundance: Rare migrant or stray along the Coastal Plain. SNA

Remarks: Periodically appears in southeastern Pennsylvania. First recorded in southeastern Pennsylvania in the late nineteenth century and again over a century later. Should be looked for in southern counties. It may become established as a breeding resident as its range expands northward.

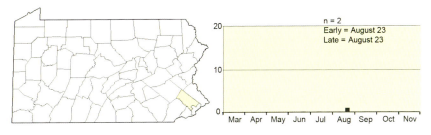

n = 2
Early = August 23
Late = August 23

♂   ♀

average wing span

Distinguishing marks: Dorsal forewing of male with yellow-orange field lacking stigma. Female forewing black with little light scaling. Ventral side unmarked tawny yellow.

Typical behavior: Males perch on low vegetation awaiting females.

Habitat: Pine barrens, sandy pine forests, grasslands with abundant little bluestem.

Larval hosts: Does not breed in Pennsylvania. Uses little bluestem (*Schizachyrium scoparium*) elsewhere.

Abundance: Rare stray. SNA

Remarks: A single stray individual was recorded in eastern Pennsylvania in 1967. This individual likely strayed from a metapopulation that ranged from New Jersey to Staten Island and Long Island. Local colonies persist in north-central New Jersey.

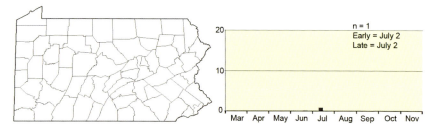

n = 1
Early = July 2
Late = July 2

♂   ♀

average wing span

Distinguishing marks: Dorsal side orange with black border. Male forewing with (1) black veins in orange patch, including end of cell bar; also lacking stigma. Female forewing with (2) prominent black scaling. Ventral uniform bright yellowish orange including fringes.

Typical behavior: Males perch on low vegetation awaiting females. Avid nectarer. Fast in flight.

Habitat: Woodland edges, old fields, wet meadows, streamsides, occasionally disturbed vacant lots.

Larval hosts: Deer tongue (*Dichanthelium clandestinum*), western panic grass (*D. acuminatum*), switchgrass (*Panicum virgatum*), plume grass (*Erianthus*), and reed canary grass (*Phalaris arundinacea*).

Abundance: Uncommon. In some localities it is found readily. S4

Remarks: One brood. Rarely, a partial second brood in fall. Overwinters as partially grown larva. Ventral side is one of the brightest of grassskippers. The similar European Skipper (p. 203) is smaller with narrower black borders and duller tan ventral side.

n = 676
Early = May 31
Late = Oct. 10

A distinctive dark bar or brand, known as the stigma, occurs on the dorsal forewings of many male skippers. The stigma consists of a collection of specially modified scales (androconia) containing pheromones. When released, pheromones attract females of the same species. The size, shape, and scale composition of the stigma are useful in distinguishing skipper species from another.

Above are examples of difficult-to-see black stigmata on dark backgrounds. Clouded Skipper (left); Little Glassywing (right).

Above are examples of stigmata of male Hesperia *that have a pale center line. Leonard's Skipper (left); Indian Skipper (right).*

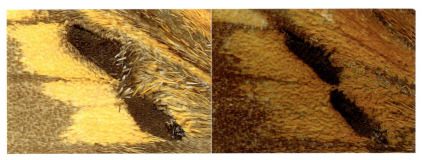

Above are examples of stigmata of male Euphyes *that have two parts. Dion Skipper (left); Two-spotted Skipper (right)*

Two of the most difficult grass skippers to distinguish in the field or in photographs are the Tawny-edged Skipper and the similar Crossline Skipper. With experience one can learn to recognize subtle differences in their stigmata.

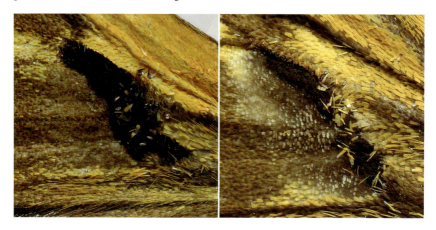

Above are examples of the thick curvilinear, S-shaped stigma of the Tawny-edged Skipper (left) and the narrower stigma of the Crossline Skipper (right). The Crossline Skipper's stigma is also less distinct, blending into a dark area.

Most stigmata consist of a single texture produced by uniform androconial scales, while others such as the Sachem's are multitextured (four or more different textures). The Sachem's stigma is large and surrounded by orange scaling, making it a dominant feature of the male's dorsal forewing. Below is an example of the multitextured stigma of the Sachem.

♂ ♀

average wing span

Distinguishing marks: Male with (1) dorsal hindwing orange patch divided by black veins; ventral hindwing with (2) extended yellow rectangle, and (3) small yellow basal spot. Female with less orange and wider black borders. Dark form female ("*pocahontas*") with prominent (4) white spots on dorsal side; ventral side purplish brown with hindwing rectangles muted.

Typical behavior: Avid nectarer. Males perch on vegetation awaiting females in open areas.

Habitat: Woodland openings and edges, damp meadows, old fields, parks, gardens.

Larval hosts: Deer tongue (*Dichanthelium clandestinum*) and bluegrass (*Poa*).

Abundance: Common throughout the state. S5

Remarks: One brood. Overwinters as partially grown larva. Dark brown female form "*pocahontas*" similar to Zabulon Skipper female, but lacks white margin on leading edge of hindwing (see following page).

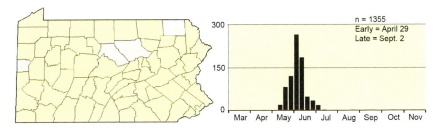

n = 1355
Early = April 29
Late = Sept. 2

The female form "*pocahontas*," on the right below, can be distinguished from a number of other female grass skippers by the presence of a white spot near the center of the forewing on the dorsal and ventral sides.

form "*pocahontas*"

Special Topic   Apiculus

The antennae of some skippers have an additional feature, the apiculus, not found in true butterflies (superfamily Papilionoidea). The apiculus is a thin distal extension of the antennal club; some of these are hooked.

Below are examples of skipper antennae, ranging from no apiculus to an example with a large apiculus. From left to right: Least Skipper, Leonard's Skipper, Common Roadside-Skipper, and Wild Indigo Duskywing.

The subspecies *viator* and *zizaniae* are found within the state. Subspecies *viator* is shown below and *zizaniae* is shown on the following page. Their differences are described in the Remarks section.

average wing span

Distinguishing marks: Larger than other resident grass skippers. Ventral hindwing with (1) one dash extending well beyond others. Male lacks stigma.

Typical behavior: Males patrol in a slow, bouncing flight through sedges and grasses looking for females.

Habitat: Marshes, swamps, wet meadows.

Larval hosts: Hairy sedge (*Carex lacustris*) in northern and northwestern inland population. Wild rice (*Zizania aquatica*) and common reed (*Phragmites australis*) in eastern and southeastern population.

Abundance: Northern and northwestern inland population (ssp. *viator*), uncommon. S2S3* Eastern and southeastern population (ssp. *zizaniae*), locally common. S4

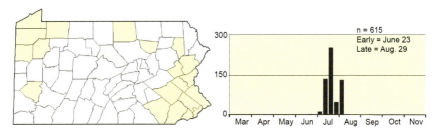

n = 615
Early = June 23
Late = Aug. 29

♂ ♀

average wing span

Subspecies zizaniae *shown above. Darker individuals of* zizaniae *shown below.*

Remarks: One brood. Overwinters as partially grown larva. The two subspecies represent an example of Great Lakes–Coastal Plain disjunction. The slightly larger eastern population (*zizaniae*) is expanding as its invasive *Phragmites* host adapts to disturbed wetlands. The key differentiating field mark of these subspecies is on the ventral hindwing. The orange ray of *viator* prominently extends from the wing base to near the outer margin; in *zizaniae* this ray is weakly defined and shorter. Two orange spots below and one above this ray are larger in *zizaniae*, even in the dark variant; in *viator* these marks are reduced and sometimes absent.

average wing span

Distinguishing marks: Dorsal forewing of male with (1) large black stigma ("black dash") surrounded by orange. Female forewing dark brown with (2) curved band of prominent white spots. Ventral side reddish brown with (3) column of smudgy yellowish-orange spots, central spot extending slightly inward.

Typical behavior: Males perch on low vegetation awaiting females.

Habitat: Marshes and wet meadows with sedges.

Larval hosts: Sedges (*Carex*), especially narrow-bladed species.

Abundance: Uncommon, but in some localities it is found readily. S3S4*

Remarks: One brood. Overwinters as partially grown larva. Adults do not wander far from hosts. Often flies with Mulberry Wing (p. 228) in sedge wetlands.

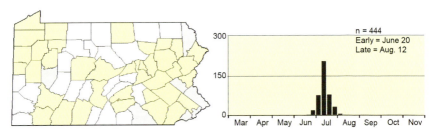

♂ ♀

average wing span

Distinguishing marks: Dorsal forewing of male with black stigma surrounded by orange patch and black border. Female forewing dark brown with curved band of light yellow spots. Dorsal hindwing with (1) small orange patch, occasionally reduced to single dash. Ventral hindwing reddish brown with (2) two long yellowish-orange rays.

Typical behavior: Males perch on low vegetation awaiting females.

Habitat: Marshes, swamps, bogs, wet meadows with sedges.

Larval hosts: Hairy sedge (*Carex lacustris*). Likely uses other sedges.

Abundance: Uncommon, locally restricted to sedge habitats. S3*

Remarks: One brood. Overwinters as partially grown larva. Adults do not wander far from hosts. Ventral hindwing similar to Broad-winged Skipper (p. 232), which has a single long ray on the ventral hindwing.

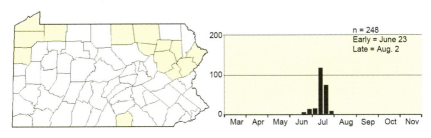

n = 248
Early = June 23
Late = Aug. 2

Two-spotted Skipper *Euphyes bimacula bimacula* (Grote and Robinson, 186[

♂   ♀

average wing span

Distinguishing marks: Dorsal forewing of male with black stigma surrounded by orange patch and black border. Female forewing dark brown with curved band of light yellow spots, two central spots largest ("two-spotted"). Ventral side dark orange with pale veins. Hindwing with white fringe, including anal margin. Underside of body light gray and palps white.

Typical behavior: Males perch on low vegetation awaiting females. Males very territorial and chase one another in spiral aerial flights.

Habitat: Marshes, swamps, bogs, wet meadows with sedges.

Larval hosts: Sedges (*Carex*).

Abundance: Uncommon, locally restricted to sedge habitats. S3*

Remarks: One brood. Overwinters as partially grown larva. Adults do not wander far from sedges. Swarthy Skipper (p. 200) is similar, but overall smaller, darker on the dorsal side, and lacks white fringe on the ventral hindwing.

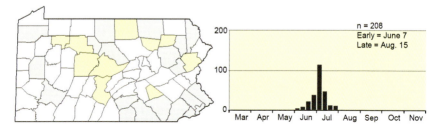

n = 208
Early = June 7
Late = Aug. 15

Dun Skipper   *Euphyes vestris metacomet*   (T. Harris, 1862)

♂   ♀

average wing span

Distinguishing marks: Dorsal side dark brown. Male forewing with faint black stigma, otherwise unmarked. Female forewing with few reduced white spots. Ventral side dark brown with purplish sheen when fresh. Hindwing usually unmarked, but some individuals with faint arc of tiny white spots.

Typical behavior: Avid nectarer. Males perch on low vegetation awaiting females.

Habitat: Woodland openings and edges, old fields, meadows, roadsides, parks, gardens.

Larval hosts: Sedges (*Carex*).

Abundance: Common throughout the state. S5

Remarks: One brood with partial second brood in some years. Overwinters as partially grown larva. Female similar to other small dark female skippers ("Three Witches") that fly at the same time, in the same habitat, and are difficult to differentiate. See Special Topic: Three Witches (p. 239).

n = 2961
Early = May 18
Late = Sept. 9

Additional examples of dorsal sides of *Euphyes* species found in Pennsylvania are shown in the left column. They are, from top to bottom: Black Dash, Dion Skipper, Two-spotted Skipper, and Dun Skipper. All photos are of males.

Shown in the right column below are two examples of ventral sides of *Euphyes* species. They are Dion Skipper (top) and Two-spotted Skipper (bottom). Note the yellowish-orange rays on the hindwing of the Dion Skipper. Also note the distinctive white fringe on anal margin of hindwing of the Two-spotted Skipper, plus its light gray body and white palps.

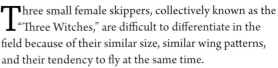

Three small female skippers, collectively known as the "Three Witches," are difficult to differentiate in the field because of their similar size, similar wing patterns, and their tendency to fly at the same time.

Close examination of postmedian spots on the dorsal forewing (shape, size, color, and position) is a useful beginning toward establishing an identification. The dorsal sides of all three species are shown on the left column at three times normal size. From top to bottom they are:

Northern Broken-Dash. Forewing spots are moderate-sized and light yellow; the most prominent two spots are generally rectangular.

Little Glassywing. Forewing spots are generally large and white; the largest spot is translucent and nearly square.

Dun Skipper. Forewing is dark with markedly reduced white spots; center spot tends to be taller than wide.

In addition to the historical "Three Witches," the dorsal sides of females of two other similar grass skippers are included in this comparison (below). They are the Tawny-edged Skipper (left) and the Crossline Skipper (right). The Tawny-edged is noticeably smaller than the other four.

Tawny-edged Skipper. Forewing has prominent orange scaling along costa; spots are generally large and more orangeish than in the other four species.

Crossline Skipper. Forewing has dull orange scaling along costa; spots are more rocket-like than rectangular as in Tawny-edged.

average wing span

Distinguishing marks: Dorsal side brown. Forewing pointed with white spots (set of three in subapical region and one to two in center). Female with additional squarish spots. Ventral side brown with extensive gray scaling ("dusted") on outer portions of wings. Hindwing with (1) white dot near base (sometimes absent).

Typical behavior: Males perch on low vegetation awaiting females. Males are strong fliers and aggressively inspect intruders.

Habitat: Dry fields, ridgetops, and right-of-ways with bluestem grasses. Often found with Cobweb Skipper, but starting flight a bit later.

Larval hosts: Big bluestem (*Andropogon gerardii*) and little bluestem (*Schizachyrium scoparium*).

Abundance: Uncommon. Concentrated in southeastern counties. S2S3*

Remarks: One brood. Overwinters as fully grown larva. Adults never wander far from hosts. The dark females of Hobomok and Zabulon Skippers (pp. 230, 229) can be mistaken for the Dusted Skipper due to gray frosting on ventral hindwings.

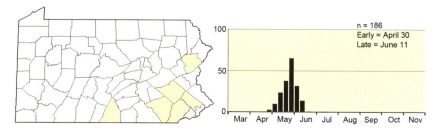

n = 186
Early = April 30
Late = June 11

♂   ♀

average wing span

Distinguishing marks: Small brown skipper with checkered margins. Forewing with curved band of white spots and subapical white spots (more prominent in female). Ventral side with extensive gray overscaling ("pepper and salt"); slight greenish sheen when fresh. Hindwing with faint arc of creamy spots.

Typical behavior: Males perch on low vegetation awaiting females.

Habitat: Woodland openings and edges, streamsides.

Larval hosts: Bearded shorthusk (*Brachyelytrum erectum*). Likely uses other grasses.

Abundance: Common. May be abundant in some localities. S4S5

Remarks: One brood. Overwinters as young larva. Similar to Common Roadside-Skipper (p. 242), but smaller overall and distinguished by arc of white spots on ventral hindwing.

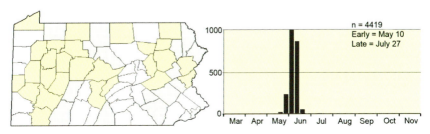

n = 4419
Early = May 10
Late = July 27

average wing span

Distinguishing marks: Brown skipper with checkered margins. Forewing with faint curved band of white spots and subapical white spots (slightly more prominent in female) and small dashes along costa. Ventral side with modest gray overscaling. Hindwing lacking arc of white spots. Small white spots along forewing of male (see magnified view) larger than in previous species.

Typical behavior: Males perch on low vegetation awaiting females.

Habitat: Woodland openings and edges, ridgetops, streamsides.

Larval hosts: Kentucky bluegrass (*Poa pratensis*), oat grass (*Avena*), bentgrass (*Agrostis*).

Abundance: Uncommon. Usually appearing when unexpected. S2S3*

Remarks: Two broods. Overwinters as fully grown larva or pupa. Similar to Pepper and Salt Skipper (p. 241), but larger with darker ventral side and lacking arc of white spots on hindwing.

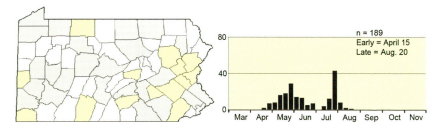

n = 189
Early = April 15
Late = Aug. 20

♂   ♀

average wing span

Distinguishing marks: Dorsal forewing with three small subapical white spots (sometimes two in male) and two larger center spots. Ventral hindwing with three distinct spots, (1) two close together ("twin spot").

Typical behavior: Avid nectarer.

Habitat: Coastal marshes.

Larval hosts: Does not breed in Pennsylvania. Primarily uses bluestem grasses in the South.

Abundance: Rare stray along Coastal Plain. SNA

Remarks: Southern resident. Periodic movements rarely along the Coastal Plain. Recorded once in state in 1960. Looks like small version of Brazilian Skipper (p. 244). The Two-spotted Skipper (*Euphyes bimacula*, p. 236) has a similar common name, but does not resemble the Twin-spot Skipper.

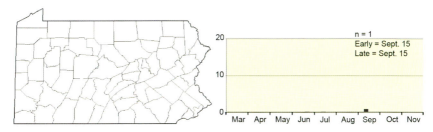

n = 1
Early = Sept. 15
Late = Sept. 15

20

10

0

Mar Apr May Jun Jul Aug Sep Oct Nov

♂ ♀

average wing span

Distinguishing marks: Long pointed forewing with row of large white spots on both sides; slight orange on leading edge. Hindwing with row of distinct white spots on both sides. Largest grass skipper in our area.

Typical behavior: Avid nectarer. Very fast flier.

Habitat: Thrives in the South where cannas grow. In Pennsylvania likely to be found nectaring in parks or gardens.

Larval hosts: Does not breed in Pennsylvania. Uses canna species (*Canna indica, C. flaccida*) in the South.

Abundance: Rare stray. SNA

Remarks: Periodic movements along Coastal Plain noted. Last recorded in state in 2012. A large robust skipper with prominent white spots should be evaluated as a possible Brazilian Skipper. Best seen while nectaring.

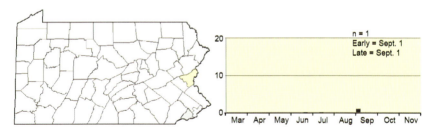

n = 1
Early = Sept. 1
Late = Sept. 1

♂   ♀

average wing span

Distinguishing marks: Forewing pointed with small pale spots on dorsal side, generally larger in female. Dorsal hindwing unmarked. Ventral side light brown with (1) pale veins on both wings. Hindwing with (2) white streak in center.

Typical behavior: Strong flier. Adults move inland to fields, roadsides, and gardens for nectar.

Habitat: Coastal salt marshes.

Larval hosts: Salt grass (*Distichlis spicata*).

Abundance: Presumed extirpated. SX*

Remarks: Two broods. Overwinters as partially grown larva. Colony existed in brackish marshes of Philadelphia and Delaware Counties on the Coastal Plain. Last recorded in 1971. Still common in coastal marshlands of southern New Jersey.

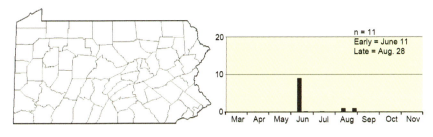

n = 11
Early = June 11
Late = Aug. 28

20

10

0

Mar Apr May Jun Jul Aug Sep Oct Nov

♂   ♀

average wing span

Distinguishing marks: Sexes similar. Forewing pointed with two white spots on dorsal side; (1) largest spot shaped like an arrowhead. Hindwing unmarked. Ventral side brown with pale veins.

Typical behavior: Avid nectarer. Strong flier.

Habitat: Woodland edges, old fields, gardens.

Larval hosts: Does not breed in Pennsylvania. Frequently uses rice cutgrass (*Leersia oryzoides*) in the South.

Abundance: Periodic migrant across the state. Wanders widely. SNA

Remarks: Typically arrives in late summer and fall. When nectaring or at rest, easily recognized by elongated forewings. See Special Topic: Fall Migrant Skippers (p. 211).

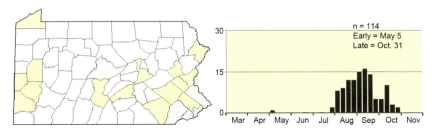

n = 114
Early = May 5
Late = Oct. 31

The number of species recorded in the state is not static; rather, it can be a dynamic situation over short periods of time measured in mere decades. Butterfly ranges expand and contract due to changing climatic conditions. New residents and visitors may turn up following systematic surveys of areas previously unexplored. Old museum collections also hold surprises; records of species "unrecorded" from the state may sit tightly in a drawer awaiting discovery. The following suppositional list provides descriptions of species that may occur in the state.

Mustard White   *Pieris oleracea oleracea*   (W. H. Edwards, 1863)

average wing span

Remarks: Resident of northern woodlands and shrubby wetlands, easily mistaken for Cabbage White and West Virginia White (pp. 54, 53). A few old records exist for New Jersey and Ohio. It should be looked for in summer (second brood) in the highest elevations of northern Pennsylvania and in old museum collections.

Seasonal dimorphism shown above: spring form (left) with grayish-green veins on ventral side; summer form, second brood (right) without vein coloration.

Large Orange Sulphur *Phoebis agarithe agarithe* (Boisduval, 1836)

♂ ♀

average wing span

Remarks: Large sulphur similar in size to Cloudless Sulphur (p. 63). Resident of the South and Southwest, and irregular migrant along Atlantic coast. Recorded in New Jersey. Look for it in late summer and fall. Males have unmarked orange dorsum. Female has yellow form, as well as cream-colored form shown above.

Mexican Yellow *Eurema mexicana mexicana* (Boisduval, 1836)

average wing span

Remarks: Resident of the Southwest and an irregular migrant. Recorded in Ohio. Look for it in western Pennsylvania in late summer and fall. Recognized by "dogface" silhouette on dorsal forewings and small pointed tail on hindwing. Compare to Southern Dogface (p. 62).

Reakirt's Blue   *Echinargus isola*   (Reakirt, [1867])

♂   ♀

average wing span

Remarks: Resident of the Southwest and an irregular migrant. Recorded in Ohio. Look for it in western Pennsylvania in late summer. Distinct row of black spots on ventral side of forewing.

Small Tortoiseshell   *Aglais urticae*   (Linnaeus, 1758)

average wing span

Remarks: A periodic introduction from Europe. May have been naturalized in New York. Look for it in the northeastern part of state in late summer and fall. Closely resembles Milbert's Tortoiseshell (p. 151). Distinguished by black dots and buff-colored bars on dorsal forewing.

White Peacock *Anartia jatrophae guantanamo* Monroe, 1942

average wing span

Remarks: Resident of the South and an irregular migrant. Recorded in New Jersey. Look for it in late summer and fall, especially after tropical storms and hurricanes. Males and females similar, white with orange and brown markings, including small eyespots.

Goatweed Leafwing *Anaea andria* (Guérin-Méneville, [1844])

average wing span

Remarks: Resident of the South and Southwest, and an irregular migrant. Recorded multiple times in Ohio. Look for it in the western part of state in late summer and fall. In September 2011, naturalist Jerry McWilliams had a brief encounter with a brilliant red-orange nymphalid at Presque Isle State Park. A photo or voucher specimen could not be obtained. The butterfly most likely was this species, migrating eastward along the southern shore of Lake Erie.

Gemmed Satyr *Cyllopsis gemma gemma* (Hübner, [1809])

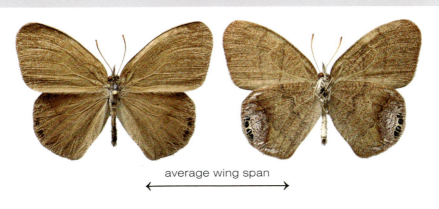

average wing span

Remarks: Resident of southern Ohio and West Virginia. Closely resembles Little Wood Satyr and Carolina Satyr (pp. 167, 166). Look for it in the southwestern part of state from late spring through summer. Males and females similar, hindwing with distinct gray patch.

Dukes' Skipper *Euphyes dukesi dukesi* (Lindsey, 1923)

♂ ♀

average wing span

Remarks: Resident of northwestern Ohio sedge wetlands. Venter similar in appearance to Dion Skipper (p. 235); dorsum is very dark. Look for it in northwestern Pennsylvania in midsummer and in old collections misidentified as *Euphyes dion*.

Eufala Skipper   *Lerodea eufala eufala*   (W. H. Edwards, 1869)

average wing span

Remarks: Resident of the South and Southwest, and an irregular migrant. Recorded in Ohio, Maryland, and New Jersey. Look for it in late summer and fall. Small grayish-brown skipper with *J*-shaped arrangement of white spots on forewings.

L isted below are the 156 species of butterflies and skippers found in Pennsylvania along with a box that can be checked when this species is first sighted, and a line for recording data such as dates and locations. Included are both the common and scientific names of each species. For scientific names *A Catalogue of the Butterflies of the United States and Canada* (Pelham 2012) is followed. In general, for common names the listing given by the North American Butterfly Association in their "Checklist of North American Butterflies Occurring North of Mexico" is followed.

Superfamily: Papilionoidea. True butterflies
Family: Papilionidae. Parnassians and Swallowtails
Subfamily: Papilioninae. Swallowtails

Pipevine Swallowtail
☐ *Battus philenor philenor*_____

Zebra Swallowtail
☐ *Eurytides marcellus*_____

Black Swallowtail
☐ *Papilio polyxenes asterius*_____

Giant Swallowtail
☐ *Papilio cresphontes*_____

Eastern Tiger Swallowtail
☐ *Papilio glaucus glaucus*_____

Canadian Tiger Swallowtail
☐ *Papilio canadensis*_____

Appalachian Tiger Swallowtail
☐ *Papilio appalachiensis*_____

Spicebush Swallowtail
☐ *Papilio troilus troilus*_____

Palamedes Swallowtail
☐ *Papilio palamedes palamedes*_____

Family: Pieridae. Whites and Sulphurs
Subfamily: Pierinae. Whites

Checkered White
☐ *Pontia protodice*_____

West Virginia White
☐ *Pieris virginiensis hyatti*_____

Cabbage White
☐ *Pieris rapae rapae*_____

Olympia Marble

☐ *Euchloe olympia*_____

Falcate Orangetip

☐ *Anthocharis midea annickae*_____

Subfamily: Coliadinae. Sulphurs

Clouded Sulphur

☐ *Colias philodice philodice*_____

Orange Sulphur

☐ *Colias eurytheme*_____

Pink-edged Sulphur

☐ *Colias interior*_____

Southern Dogface

☐ *Zerene cesonia cesonia*_____

Cloudless Sulphur

☐ *Phoebis sennae eubule*_____

Orange-barred Sulphur

☐ *Phoebis philea philea*_____

Sleepy Orange

☐ *Abaeis nicippe*_____

Little Yellow

☐ *Pyrisitia lisa lisa*_____

Dainty Sulphur

☐ *Nathalis iole iole*_____

Family: Lycaenidae. Gossamer Wings

Subfamily: Miletinae. Harvesters

Harvester

☐ *Feniseca tarquinius tarquinius*_____

Subfamily: Lycaeninae. Coppers

American Copper

☐ *Lycaena phlaeas hypophlaeas*_____

Bronze Copper

☐ *Lycaena hyllus*_____

Bog Copper

☐ *Lycaena epixanthe michiganensis*_____

Subfamily: Theclinae. Hairstreaks

Great Purple Hairstreak

☐ *Atlides halesus halesus*_____

Coral Hairstreak

☐ *Satyrium titus titus*_____

Acadian Hairstreak

☐ *Satyrium acadica acadica*_____

Edwards' Hairstreak

☐ *Satyrium edwardsii edwardsii*_____

Banded Hairstreak

☐ *Satyrium calanus falacer*_____

Hickory Hairstreak

☐ *Satyrium caryaevorus*_____

Striped Hairstreak

☐ *Satyrium liparops strigosa*_____

Oak Hairstreak

☐ *Satyrium favonius ontario*_____

Brown Elfin

☐ *Callophrys augustinus croesioides*_____

Hoary Elfin

☐ *Callophrys polios polios*_____

Frosted Elfin

☐ *Callophrys irus irus*_____

Henry's Elfin

☐ *Callophrys henrici henrici*_____

Eastern Pine Elfin

☐ *Callophrys niphon niphon*_____

Juniper Hairstreak

☐ *Callophrys gryneus gryneus*_____

White M Hairstreak

☐ *Parrhasius m-album*_____

Gray Hairstreak

☐ *Strymon melinus humuli*_____

Red-banded Hairstreak

☐ *Calycopis cecrops*_____

Early Hairstreak

☐ *Erora laeta*_____

Subfamily: Polyommatinae. Blues

Marine Blue
☐ *Leptotes marina*_____

Eastern Tailed-Blue
☐ *Cupido comyntas comyntas*_____

Spring Azure
☐ *Celastrina ladon* _____

Northern Spring Azure
☐ *Celastrina lucia lucia*_____

Summer Azure
☐ *Celastrina neglecta* _____

Cherry Gall Azure
☐ *Celastrina serotina* _____

Appalachian Azure
☐ *Celastrina neglectamajor* _____

Dusky Azure
☐ *Celastrina nigra* _____

Silvery Blue
☐ *Glaucopsyche lygdamus nittanyensis*_____

Karner Blue
☐ *Plebejus samuelis* _____

Family: Riodinidae. Metalmarks

Subfamily: Riodininae. Metalmarks

Northern Metalmark
☐ *Calephelis borealis*_____

Swamp Metalmark
☐ *Calephelis muticum*_____

Family: Nymphalidae. Brushfoots

Subfamily: Libytheinae. Snouts

American Snout
☐ *Libytheana carinenta bachmanii*_____

Subfamily: Danainae. Milkweed Butterflies

Monarch
☐ *Danaus plexippus plexippus*_____

Queen
☐ *Danaus gilippus berenice*_____

Subfamily: Heliconiinae. Longwings and Fritillaries

Gulf Fritillary
☐ *Agraulis vanillae nigrior* _____

Variegated Fritillary
☐ *Euptoieta claudia claudia* _____

Great Spangled Fritillary
☐ *Speyeria cybele cybele* _____

Aphrodite Fritillary
☐ *Speyeria aphrodite aphrodite* _____

Atlantis Fritillary
☐ *Speyeria atlantis atlantis* _____

Regal Fritillary
☐ *Speyeria idalia idalia* _____

Diana Fritillary
☐ *Speyeria diana* _____

Silver-bordered Fritillary
☐ *Boloria selene myrina* _____

Meadow Fritillary
☐ *Boloria bellona bellona* _____

Subfamily: Nymphalinae. True Brushfoots

Gorgone Checkerspot
☐ *Chlosyne gorgone carlota* _____

Silvery Checkerspot
☐ *Chlosyne nycteis nycteis* _____

Harris' Checkerspot
☐ *Chlosyne harrisii liggetti* _____

Baltimore Checkerspot
☐ *Euphydryas phaeton phaeton* _____

Pearl Crescent
☐ *Phyciodes tharos tharos* _____

Northern Crescent
☐ *Phyciodes cocyta selenis* _____

Tawny Crescent
☐ *Phyciodes batesii batesii* _____

Common Buckeye
☐ *Junonia coenia coenia* _____

Question Mark
☐ *Polygonia interrogationis* _____

Eastern Comma
☐ *Polygonia comma* _____

Green Comma
☐ *Polygonia faunus faunus* _____

Gray Comma
☐ *Polygonia progne* _____

Compton Tortoiseshell
☐ *Nymphalis l-album j-album* _____

California Tortoiseshell
☐ *Nymphalis californica* _____

Mourning Cloak
☐ *Nymphalis antiopa antiopa* _____

Milbert's Tortoiseshell
☐ *Aglais milberti milberti* _____

American Lady
☐ *Vanessa virginiensis* _____

Painted Lady
☐ *Vanessa cardui* _____

Red Admiral
☐ *Vanessa atalanta rubria* _____

Subfamily: Limenitidinae. Admirals and Relatives

White Admiral
☐ *Limenitis arthemis arthemis* _____

Red-spotted Purple
☐ *Limenitis arthemis astyanax* _____

Viceroy
☐ *Limenitis archippus archippus* _____

Subfamily: Apaturinae. Emperors

Hackberry Emperor
☐ *Asterocampa celtis celtis* _____

Tawny Emperor
☐ *Asterocampa clyton clyton* _____

Subfamily: Satyrinae. Satyrs and Browns

Northern Pearly-Eye
☐ *Lethe anthedon anthedon* _____

Eyed Brown
☐ *Lethe eurydice eurydice*_____

Appalachian Brown
☐ *Lethe appalachia leeuwi*_____

Carolina Satyr
☐ *Hermeuptychia sosybius*_____

Little Wood Satyr
☐ *Megisto cymela cymela*_____

Common Ringlet
☐ *Coenonympha tullia inornata*_____

Common Wood Nymph
☐ *Cercyonis pegala nephele*_____
☐ *Cercyonis pegala alope*_____

Superfamily: Hesperioidea, Skippers
Family: Hesperiidae. Skippers
Subfamily: Eudaminae. Dicot Skippers

Silver-spotted Skipper
☐ *Epargyreus clarus clarus*_____

Long-tailed Skipper
☐ *Urbanus proteus proteus*_____

Golden-banded Skipper
☐ *Autochton cellus*_____

Hoary Edge
☐ *Achalarus lyciades*_____

Northern Cloudywing
☐ *Thorybes pylades pylades*_____

Southern Cloudywing
☐ *Thorybes bathyllus*_____

Confused Cloudywing
☐ *Thorybes confusis*_____

Subfamily: Pyrginae. Spread-Winged Skippers

Dreamy Duskywing
☐ *Erynnis icelus*_____

Sleepy Duskywing
☐ *Erynnis brizo brizo*_____

Juvenal's Duskywing

☐ *Erynnis juvenalis juvenalis*_____

Horace's Duskywing

☐ *Erynnis horatius* _____

Mottled Duskywing

☐ *Erynnis martialis* _____

Zarucco Duskywing

☐ *Erynnis zarucco*_____

Funereal Duskywing

☐ *Erynnis funeralis* _____

Columbine Duskywing

☐ *Erynnis lucilius* _____

Wild Indigo Duskywing

☐ *Erynnis baptisiae* _____

Persius Duskywing

☐ *Erynnis persius persius*_____

Hayhurst's Scallopwing

☐ *Staphylus hayhurstii* _____

Appalachian Grizzled Skipper

☐ *Pyrgus wyandot*_____

Common Checkered-Skipper

☐ *Pyrgus communis communis*_____

Common Sootywing

☐ *Pholisora catullus* _____

Subfamily: Heteropterinae. Skipperlings

Arctic Skipper

☐ *Carterocephalus palaemon mandan*_____

Subfamily: Hesperiinae. Grass Skippers

Swarthy Skipper

☐ *Nastra lherminier*_____

Clouded Skipper

☐ *Lerema accius* _____

Least Skipper

☐ *Ancyloxypha numitor*_____

European Skipper

☐ *Thymelicus lineola lineola*_____

Fiery Skipper
☐ *Hylephila phyleus phyleus*_____

Sachem
☐ *Atalopedes campestris huron*_____

Leonard's Skipper
☐ *Hesperia leonardus leonardus*_____

Cobweb Skipper
☐ *Hesperia metea metea*_____

Dotted Skipper
☐ *Hesperia attalus slossonae*_____

Indian Skipper
☐ *Hesperia sassacus sassacus*_____

Peck's Skipper
☐ *Polites peckius peckius*_____

Tawny-edged Skipper
☐ *Polites themistocles themistocles*_____

Crossline Skipper
☐ *Polites origenes origenes*_____

Long Dash
☐ *Polites mystic mystic*_____

Whirlabout
☐ *Polites vibex vibex*_____

Little Glassywing
☐ *Pompeius verna verna*_____

Northern Broken-Dash
☐ *Wallengrenia egeremet*_____

Southern Broken-Dash
☐ *Wallengrenia otho otho*_____

Arogos Skipper
☐ *Atrytone arogos arogos*_____

Delaware Skipper
☐ *Anatrytone logan logan*_____

Mulberry Wing
☐ *Poanes massasoit massasoit*_____

Zabulon Skipper
☐ *Poanes zabulon*_____

Hobomok Skipper
☐ *Poanes hobomok hobomok*_____

Broad-winged Skipper

☐ *Poanes viator viator*_____

☐ *Poanes viator zizaniae* _____

Black Dash

☐ *Euphyes conspicua orono*_____

Dion Skipper

☐ *Euphyes dion* _____

Two-spotted Skipper

☐ *Euphyes bimacula bimacula*_____

Dun Skipper

☐ *Euphyes vestris metacomet*_____

Dusted Skipper

☐ *Atrytonopsis hianna hianna*_____

Pepper and Salt Skipper

☐ *Amblyscirtes hegon* _____

Common Roadside-Skipper

☐ *Amblyscirtes vialis*_____

Twin-spot Skipper

☐ *Oligoria maculata* _____

Brazilian Skipper

☐ *Calpodes ethlius* _____

Salt Marsh Skipper

☐ *Panoquina panoquin* _____

Ocola Skipper

☐ *Panoquina ocola ocola* _____

A detailed history of butterfly explorations and the development of lepidopterology in Pennsylvania has yet to be written. The literature is scattered and challenging to assemble. Significant portions of it are covered here and lay the groundwork for the future.

Precolonial Period (1000–1600)

Prior to contact with Europeans, Pennsylvania was a pristine wilderness inhabited by Native Americans. To them this land was an ancient, sacred place. The earliest European explorers in eastern Pennsylvania met the peaceful Lenni-Lenape (or Delaware), whose villages were located near the Delaware River. For the Delaware, butterflies (*memekas*) were part of their spiritual life and often were depicted in ceremonial dances.

Colonial Period (1600–1776)

In 1681, William Penn (1644–1718) was granted a charter for the Pennsylvania Colony. In letters to friends in London, Penn described the beauty of the forested land with its outstanding trees, flowers, and animals. There is no record that Penn or his followers sent insects back to England for scientific description; the intense flurry to name plants and animals from the New World would not begin for another half century. Swedish naturalist Peter Kalm (1716–1779) arrived in Philadelphia in 1748. For three years he explored Pennsylvania, New Jersey, New York, and eastern Canada. Upon his return to Sweden in 1751, he shared his collection of plants and butterflies with Carl von Linne (Linnaeus) (1707–1778), founder of the binomial system of modern taxonomy (genus and species). In the tenth edition of his *Systema Naturae* (1758), Linnaeus became the first to officially name American insects, including the Monarch (*Danaus plexippus*), for which he acknowledged Kalm as the collector. The original specimens that Linnaeus used in his description of the Monarch were collected in Pennsylvania and New York. For the remainder of the eighteenth century, collecting mania intensified and prominent European entomologists—e.g., Pieter Cramer, Dru Drury, Johan C. Fabricius—continued to name hundreds of new American insects.

First Pennsylvania Period (1776–1840): Birth of a Nation, Earliest Museums, and First American Entomologists

The Peale Museum opened in Philadelphia in 1786. Founded by master American painter Charles Willson Peale (1741–1827), the museum displayed art and curiosities of natural history to the general public and continued to do so for over fifty years. Ten years after the museum's founding, in 1796, Charles's son Titian Ramsay Peale I (1780–1798) authored the first American book on insects, *Drawings of American Insects; Shewing Them in Their Several States*. This early work featured hand-colored sketches of the Black Swallowtail (*Papilio polyxenes*), Monarch (*Danaus plexippus*), Question Mark (*Polygonia interrogationis*), Mourning Cloak (*Nymphalis antiopa*), and many moths observed in the Philadelphia area. In 1806 Frederick Valentine Melsheimer (1749–1814) published *A Catalogue of Insects of Pennsylvania*, volume one of an intended three-volume work. Illness prevented the publication of the remaining works, however. In 1812 the Academy of Natural Sciences of Philadelphia was founded "for the encouragement and cultivation of the sciences" (ANSP Charter). It is the oldest such institution in the Western Hemisphere. Shortly after its founding, prestigious members of the Academy began to collect, catalogue, and publish findings from local habitats and faraway explorations of the continent.

From 1824 to 1828, Thomas Say (1787–1834), born in Philadelphia, a self-taught naturalist and regarded as "Father of American Entomology," published his *American Entomology, or Descriptions of Insects of North America* in three volumes. The work was the first of its kind in the United States and featured illustrations of several butterflies from the Philadelphia area by Titian Ramsay Peale II (1799–1885). The youngest son of Charles Willson Peale, Titian was given the name of his older brother, who died in the yellow fever epidemic of 1798. The younger Titian's talents as a naturalist and scientific illustrator were acclaimed even when he was a teenager. In addition to illustrating for Say's *American Entomology*, he published his own *Lepidoptera Americana*, volume 1, number 1, in 1833—the first American publication dedicated to Lepidoptera. This brief volume, accompanied by a prospectus, was an inspiration to future American lepidopterists. Though it was discontinued, *Lepidoptera Americana* became the earliest phase of a much larger manuscript by Peale, *The Butterflies of North America*. Although unfortunately this work was never published during his lifetime, the manuscript was recently discovered and published in 2015 by the American Museum of Natural History (see book review by Calhoun and Wright 2016). It features numerous butterfly illustrations from material collected and reared in the Philadelphia area. Peale's large collection of butterflies currently resides in the Academy of Natural Sciences of Philadelphia.

This period continued to see European publications describe new American butterflies; some were captured in or near Philadelphia, e.g., Banded Hairstreak (*Satyrium calanus falacer*) and Milbert's Tortoiseshell (*Aglais milberti*) in Godart's *Encylopédie méthodique* (1819–1824) and Compton Tortoiseshell (*Nymphalis l-album j-album*) in Jean Alphonse Boisduval and John Le Conte's *Histoire générale et iconographie des lépidoptères et des chenilles de l'Amérique septentrionale* (1827–[1837]). Thomas Say, a strong advocate of American science, insisted that the quality of the nation's scientists was equivalent to that of those in Europe and he urged stateside entomologists to study and name their own insects.

Second Pennsylvania Period (1840–1910): Earliest Entomological Societies, the Next Wave of Museums, and a New Wave of Authors

The first society devoted exclusively to American entomology, the Entomological Society of Pennsylvania (ESP), was founded in 1842 in York. It was soon followed in 1859 by the establishment of the Entomological Society of Philadelphia, which changed its name to the American Entomological Society (AES) in 1867. Both ESP and AES remain active today. In 1855, the Wagner Free Institute of Science opened in Philadelphia; the institute moved to a new permanent hall in 1865 and featured an insect collection among its extensive holdings. In 1862, at the October 13 monthly meeting of the Entomological Society of Philadelphia, James Ridings (1803–1880) communicated that he had taken a black female *Papilio glaucus* in connection with a yellow male *Papilio turnus* as far back as 1832. In 1863, at the January 12th monthly meeting of the Entomological Society of Philadelphia,

Benjamin D. Walsh (1808–1869) provided evidence that *Papilio glaucus* and *Papilio turnus* are identical. This was the first confirmation of sexual dimorphism in swallowtails.

In 1867 August Radcliffe Grote (1841–1903) and Coleman T. Robinson (1838–1872) described two new butterfly species from Philadelphia, the Two-spotted Skipper (*Euphyes bimacula*) and Henry's Elfin (*Callophrys henrici*). In the following year, William Henry Edwards (1822–1909) began the first installments of his *Butterflies of North America* (1868–1872), one of the most important entomological publications of the nineteenth century. His primary illustrator, artist Mary Peart (1837–1917), and his publisher, AES, both were located in Philadelphia. In 1869, Simon Snyder Rathvon (1812–1891) published a list of butterflies of Lancaster County, the first faunal county list to appear in the state (53 species). Famous sculptor Herman Strecker (1836–1901) of Reading amassed a large collection of American and foreign Lepidoptera. From 1872 to 1878, he published the multivolume *Lepidoptera, Rhopaloceres and Heteroceres, Indigenous and Exotic*, which contained recognizable figures of Pennsylvania species.

In 1884, Edward Dempster Merrick (1832–1911), a collector of art and butterflies, opened the Merrick Art Gallery in New Brighton. The gallery was expanded in 1903 to include the Merrick Museum featuring Lepidoptera and other curios. The museum was managed by Edward's brother Frank Angelo Merrick (1845–1912) and his grandson Harry Duncan Merrick (1868–1907), both of whom were avid collectors of butterflies and moths of Beaver County, active correspondents, and authors of entomological papers. The museum dispersed in 1911, and its Lepidoptera collection is now held in the Smithsonian National Museum of Natural History. In 1889, Henry Skinner (1861–1926) and Eugene Murray Aaron (1852–1940) published *A List of Butterflies of Philadelphia, Pa.*, the first formal and detailed annotated list (86 Pennsylvania species) of a region where collecting had commenced nearly a century before.

The Carnegie Museum of Natural History was founded in Pittsburgh by industrialist Andrew Carnegie in 1895. The director of the museum, William Jacob Holland (1848–1932), former chancellor of the Western University of Pennsylvania (now the University of Pittsburgh), published *The Butterfly Book* (1898), which featured Pennsylvania specimens and types from the W. H. Edwards Collection that the museum had acquired. Meetings of the Entomological Society of Western Pennsylvania, founded in 1902 and with a membership that included many prominent entomologists, routinely convened at the Carnegie Museum.

Starting in 1903, under the direction of Harvey A. Surface (1867–1941), the Pennsylvania Department of Agriculture developed a significant insect collection over the next fifty years, featuring many butterflies. In 1905 the State Museum of Pennsylvania was founded for "the preservation of objects illustrating the flora and fauna of the state, and its mineralogy, geology, architecture, arts and history" (Commonwealth Law, 1905, Act 43); the zoology division moved into the museum in Harrisburg in 1907. That same year, the Reading Public Museum opened—this museum was largely based on the extensive personal collection of Levi W. Mengel (1868–1941) and featured high-quality paintings, American Indian relics, and a large collection of Lepidoptera; it eventually moved to a permanent home in 1928. In 1908, Henry Engel (1873–1943) published *A Preliminary List of the Lepidoptera of Western Pennsylvania Collected in the Vicinity of Pittsburgh*, detailing 78 species. The Everhart Museum, featuring art and a natural history collection, was founded in Scranton in 1908.

Third Pennsylvania Period (1910–1980): Avid Period of Collecting, Popular Butterfly Books for the Public, and More Detailed Faunal Surveys of State

In 1911 Reverend J. C. Stamm of Danville published *Butterflies of Montour County* (61 species), including specimens from the collection of Hugh Bradshaw Meredith (1853–1929). In 1915 Reuben Nelson Davis (1858–1934), curator of the Everhart Museum, published an *Illustrated and Annotated Catalog of the Butterflies of Lackawanna County*, detailing 82 species, which was largely based on prolific collection and notes of lepidopterist Max Rothke (1868–1936) of Scranton. When Rothke died, his collection of 20,000 butterflies and moths was purchased by Cyril dos Passos (1887–1986) and donated to the American Museum of Natural History in New York City. In 1931 Holland published a revised edition of *The Butterfly Book*, including several interesting Pennsylvania records, such as the Dion Skipper (*Euphyes dion*) in western Pennsylvania (Crawford County). A decade later, in 1941, Roswell Carter Williams Jr. (1869–1946) published *A List of Butterflies which May Be Found within 50 Miles of Philadelphia (Lepid.: Rhopalocera)* (120 Pennsylvania species).

The Lepidopterists' Society was founded in 1947 to "promote the scientifically sound and progressive study of Lepidoptera" (constitution of the Lepidopterists' Society). A journal and season summary dispersed information about state butterfly fauna and ranges; Pennsylvania contributors participated from the beginning. In 1951 Alexander Barrett Klots (1903–1989) published *A Field Guide to the Butterflies of North America East of the Great Plains*. This was the first true "field" guide of eastern butterflies; it depicted a few Pennsylvania specimens on color plates. The first statewide Pennsylvania list of butterflies (117 species) and moths, *The Lepidoptera of Pennsylvania: A Manual*, by Harrison Morton Tietz (1895–1963), was published in 1952. Records indicate that the data in the manual did not extend past 1944.

The North Museum of Natural History and Science was founded by Franklin & Marshall College in Lancaster in 1953; the museum collections traced their existence back to the nineteenth century. In 1966, Arthur M. Shapiro (1946–) published *Butterflies of the Delaware Valley*, the first annotated publication on Philadelphia area butterflies since the work of Skinner and Aaron (1889). This work provided new information on flight dates, larval food plants, and county records on 120 Pennsylvania species. In 1969 the Frost Entomological Museum opened on the campus of Penn State University in State College; its holdings were transferred from the university's entomology department and contain late nineteenth- and early twentieth-century material.

Between 1927 and 1963, Pennsylvania-born collectors Franklin Hugo Chermock (1906–1967) and his brother Ralph Lucien Chermock (1918–1977) described over 50 taxa of Lepidoptera. Frank Chermock's collection of 56,000 specimens presently resides in the McGuire Center for Lepidoptera and Biodiversity in Gainesville, Florida. Ralph Chermock continued collecting and describing Lepidoptera after the death of his brother in 1967 until his own death ten years later. His collection of 30,000 specimens is now in the Alabama Museum of Natural History on the campus of University of Alabama in Tuscaloosa. Another Pennsylvania-based collector, Harry Kendon Clench (1925–1979), served as president of the Lepidopterists' Society and associate curator of entomology at Carnegie Museum of Natural History. Clench published seventeen papers on Pennsylvania, including important surveys of the museum's field station, the Powdermill Nature Reserve in Westmoreland County. Upon his death in 1979, his collection was given to the Carnegie Museum, where it remains today.

Fourth Pennsylvania Period (1980–Present): Advent of Butterfly-Watching and Photography, Butterflies as Indicator Species in Conservation Activities, and Dissemination of Data via Internet

A new era of inventory and conservation began. In 1984, Robert Michael Pyle (1947–) published *The Audubon Society Handbook for Butterfly Watchers* and initiated the importance of observation and recording changes in butterfly habitats and ranges. In 1985, Paul A. Opler (1938–) summarized the status of 16 Pennsylvania butterfly species of special concern; it completed the first phase of a coordinated biological inventory by the Pennsylvania Biological Survey (PBS), which was formed in 1979. In 1989, the lepidopterist community lost George Escott Ehle (1913–1989), a longtime resident of and avid collector in Lancaster County. His annual contributions to the *Season Summary of the Lepidopterists' Society* and willingness to share information with others were legendary, and his collection and unpublished Lancaster County butterfly checklist (99 species) are now held in the North Museum on the campus of Franklin & Marshall College in Lancaster.

In the late 1980s the last remaining population of Regal Fritillary (*Speyeria idalia*) in Pennsylvania became restricted to the Fort Indiantown Gap (FIG) National Guard Training Center. In 1993 Jeffrey Glassberg published *Butterflies through Binoculars: A Field Guide to Butterflies of the Boston–New York–Washington Region*, which included a local checklist of butterflies at John Heinz National Wildlife Refuge at Tinicum in Philadelphia. In 1995 Opler published detailed range maps of butterflies of the eastern United States; these maps were privately distributed as early as 1983 and led to the birth of mapping county-by-county distributions. Also in 1995, David M. Wright distributed the first edition of *Atlas of Pennsylvania Butterflies*, featuring county distribution maps of 147 species; this total expanded to 156 species in the fourteenth edition twenty years later. In 2002 the Oakes Museum of Natural History opened on the campus of Messiah College in Mechanicsburg, featuring important collections of local butterfly collectors.

After the start of the new millennium, citizen scientists began to avidly report their observations and photos to Internet sources and the Pennsylvania Natural Heritage Program, a collective effort that rapidly disseminates information regarding butterfly population fluctuations and movements. Several species new to Pennsylvania were first discovered in this time period during fieldwork conducted by dedicated amateur lepidopterists—for example, the Common Ringlet (*Coenonympha tullia*), Carolina Satyr (*Hermeuptychia s osybius*), Appalachian Tiger Swallowtail (*Papilio appalachiensis*), and Cherry Gall Azure (*Celastrina serotina*).

abdomen: Third or terminal body part.

adult: Final and most conspicuous stage of the butterfly life cycle. Also known as imago.

anal angle: Junction of the inner and outer margins of the hindwing.

androconia: Modified wing scales on male butterflies that release pheromones in courtship.

antennae: Paired filamentous appendages on the butterfly head, bearing clubbed sensory organs at the tips.

apex: Area at the tip of the forewing.

apiculus: Hooked extension of the antennal club in skippers.

aposematic: Having bright warning colors that advertise unpalatability to potential predators.

band: Colored stripe on adult wings or body.

basal: In the area near the base of the wing, closest to the body.

basking: Activity in which butterflies orient themselves to gather solar radiation.

Batesian mimicry: Type of mimicry in which a palatable species (mimic) resembles an unpalatable species (model). Compare to **Müllerian mimicry**.

binomial nomenclature: Scientific method of naming each species with a two-part Latinized name, the first indicating the genus and the second being the species name.

bivoltine: Having two flights per year.

blend zone: Area where two different populations (usually species, subspecies, or ecotypes) meet and interbreed, creating hybrids.

boreal: Of the northern coniferous forest region.

brood: A single generation of adult butterflies that emerge more or less synchronously.

bursa copulatrix: Pouch or chamber in the female abdomen designed to receive male gametes.

byre : Shelter at base of a host tree where ants tend larvae (see Edwards' Hairstreak).

canopy species Species that reside primarily in the upper layer of forest.

caterpillar: Wormlike immature stage of Lepidoptera. Also known as the **larva**.

cell: Wing area enclosed by veins, usually used in reference to the **discal cell**.

chevron: *V*-shaped mark.

chrysalis: Third stage of butterfly life cycle, in which it develops a hard case and transforms into an adult. Also known as the **pupa**.

claspers: Paired structure at the end of the male abdomen designed to hold a female during mating. Also known as **valves**.

club: Sensory organ at the tip of the antenna.

coevolution: Parallel evolution of two kinds of organisms where any change results in an adaptive response in the other.

complex: See **species complex**.

compound eye: An eye consisting of an array of smaller visual units (ommatidia).

costa: Leading or forward edge of a wing.

costal fold: Flap on the leading edge of the male forewing containing scent scales.

cremaster: Hooks on the tail end of the chrysalis, used to anchor on a silken pad.

crypsis: Disguising coloration or camouflage, making the subject hard to see.

cuticle: The exoskeleton and epidermal cells that secrete it.

diapause: State of arrested development in which a butterfly passes unfavorable seasons.

dimorphism: Occurrence of two distinct forms within the same species population.

discal cell: Cell that extends from the wing base to a cross-vein in the middle of the wing.

diversity: Number and variety of butterflies known to exist in a given area.

dorsum: Upper surface of wings and body.

ecotype: A population distinct from other related taxa, due to different environmental associations. May or may not have a taxonomic name.

eclosion: Emergence of adult butterfly from the chrysalis.

ecosystem: A community of living organisms in conjunction with the nonliving components of the environment (air, water, soil).

egg: Initial life stage of a butterfly. Placed by adult females on plants suitable for the next stage (**larva**).

estivate: To spend a portion of the summer in an inactive state.

exoskeleton: Hard external body layer, consisting primarily of a water-repellent chemical called chitin.

extinct: Taxon with no living representatives, as in a species that is no longer exists.

extirpated: Taxon with no individuals surviving in a given area which it formerly occurred, as in a species that no longer resides in state, but still exists elsewhere in its range.

eyespot: Concentric decorative circle on wing, resembling a mammalian eye.

falcate: Forewing with a hooked tip.

family: Taxonomic rank between order (Lepidoptera) and genus in the Linnaean system. Six butterfly families are recognized in North America.

fauna: Animals of a region or given place. Can also be specific animals (butterfly fauna).

flight: A single generation of adults.

flight period: Time of year when adults of a given species can be found.

forewings: Forward or front pair of wings.

form: Term referring to one variety within a phenotypically variable species.

fringe: Thin, outward-projecting scales on the outer border of both wings.

generation: All the individuals of a complete life cycle (egg to adult).

genitalia: Reproductive organs of adult butterflies, located in last abdominal segments.

genus: Taxonomic rank between family and species in the Linnaean system.

head: First or anterior body part. The head contains the antennae, eyes, mouthparts, and a rudimentary brain.

hibernaculum: Winter nest or shelter.

hibernation: Overwintering in an inactive state.

hindwings: Rear or posterior pair of wings.

honeydew: Sweet liquid secreted by some insects and insect galls, often actively tended by ants. Sometimes used by adult butterflies as a nutrition source.

hyaline spots: Glasslike, translucent spots on the wings of some butterflies.

hybrid: Offspring resulting from the interbreeding of two different species, subspecies, or ecotypes (forms).

hybrid zone: Area of overlap between two species, subspecies, or ecotypes (forms) where hybridization takes place.

hybridize: Mixing through sexual reproduction of two animals of different species, subspecies, or ecotypes.

inner margin: Trailing or hind edge of the forewing.

instar: Name for larva between each larval molt. There may be four to six instars in the complete larval stage.

intergrade: See **hybrid**.

introduced: Not native to a region.

iridescence: Brilliant luminous appearance that seems to change color when seen at different angles.

irruption: Sudden increase in a butterfly population, usually from a migrating invasion.

labial palps: Paired sensory structures found on each side of the proboscis.

larva: Second stage of a butterfly's life cycle. Also known as the **caterpillar**.

larval hostplant: Any plant eaten by larva and on which the eggs are normally laid. Also known as foodplant.

Lepidoptera: Order of insects comprising butterflies and moths. Second-largest order of insects.

local: Found in small colonies or restricted to a very specific habitat.

marginal: Wing area just inside the outer edge.

mate locating: Behavior bringing the sexes together. See **patrolling** and **perching**.

median: Central or middle portion of the wing, halfway between the base and apex.

mesothorax: Second or middle segment of the thorax.

metamorphosis: Process of transforming from an immature to adult form in distinct stages.

metapopulation: Group of local populations connected by an occasional exchange of individuals.

metathorax: Third or last segment of the thorax.

migrant: Butterfly that makes long-distance flights. May be from a regular seasonal occurrence or an irregular sporadic event.

mimic: Adult butterfly that resembles or imitates the appearance of another.

mimicry: The resemblance of two or more unrelated butterflies, adapted to benefit one or all. See **Batesian mimicry** and **Müllerian mimicry**.

mimicry ring: Group of species mimicking the same pattern. A ring usually has a mainstay of unpalatable models, where each species acts as both a mimic and a model. It may also include one unpalatable model and several Batesian mimics.

model: Adult butterfly that serves as example for others to imitate.

molt: To shed the larval skin or exoskeleton.

morph: Any individual form or variety found within a polymorphic species. See **polymorphism**.

morphology: Form and structure of an animal.

Müllerian mimicry: Type of mimicry in which two or more unpalatable species display a very similar appearance (phenotype), which allows predators to learn to avoid the phenotype. Compare to **Batesian mimicry**.

multivoltine: Having three or more flights per year.

nectaring: Act of feeding on nectar, a sweet liquid produced by flowers to attract insects.

nest: Structure of leaves and silk made by larva for concealment when not feeding.

nomenclature: System of names and rules for forming these terms. See **binomial nomenclature**.

nudum: Scaleless area on the underside of the antennal tip.

ommatidium: Each individual simple eye or facet of the compound eye of adult butterfly.

outer margin: Outermost edge of the membranous wings.

overwinter: To pass through cold winter conditions in an arrested metabolic state. Butterflies may overwinter as an egg, larva, pupa, or adult.

oviposit: To lay one or more eggs.

ovum (pl. ova): Initial life stage of a butterfly. See **egg**.

palps: Paired sensory structures found on each side of the proboscis. See **labial palps**.

parasitoid: Free-living adult insect (wasp, fly) whose larvae feed on or within a host. Parasitoids are important natural enemies of butterfly eggs, larvae, and pupae.

partial brood: Group of adults that emerged from pupation while most remained in diapause.

patrolling: Mate-locating strategy in which male butterflies continuously fly in a specific territory seeking receptive females.

perching: Mate-locating strategy in which male butterflies rest on a stationary object awaiting receptive females.

phenogram: Diagram showing relationships, such as adult flight period (temporal distribution).

phenology: Study of the timing of biological events, such as adult butterfly flight periods.

phenotype: Appearance of an organism, particularly, in the case of butterflies, with reference to wing colors and patterns.

pheromone: Chemical that induces a specific behavioral change, as in sexual courtship.

polymorphism: Occurrence of several forms within a species. Usually restricted to forms under genetic control, such as sexual forms. Compare to **polyphenism**.

polyphenism: Occurrence of several forms within a species. Usually restricted to forms induced by environmental influences, like seasonal forms. Compare to **polymorphism**.

population: Group of butterflies belonging to the same species and occupying a defined area.

postbasal: Relating to the area of the wing lying just beyond the base.

postbasal band: Band or stripe located between the basal and median area of wings.

postmedian: In the area of the wing lying just beyond the median or central area.

proboscis: Coiled tongue of the adult butterfly, used to siphon liquids like nectar and soil moisture. The proboscis is kept coiled beneath the head when not in use.

prolegs: Small fleshy legs on the abdomen of the larva.

prothorax: First, or anterior, segment of the thorax.

pterin: Class of chemical compounds responsible for orange, yellow, and white coloration.

puddle: To siphon moisture from soils; a method of harvesting mineral salts (to mineralize).

puddle party: Aggregation of butterflies at mud, usually consisting of males.

pupa: Third stage of butterfly life cycle in which it develops a hard case and transforms into an adult. Also known as the **chrysalis**.

pupate: To transform into a pupa or chrysalis.

range: Geographic area occupied by a species or subspecies.

rejection behavior: Behavior performed by an unreceptive female to discourage a male. This often involves wing fluttering or raising the abdomen.

relict: Population left behind the main range of a species, such as following a glacial period.

scales: Small, flattened, modified hairs covering the wings and bodies of Lepidoptera. Scale colors are produced through chemical pigments, structural iridescence, or both.

sexual dimorphism: Conspicuous phenotypic differences between the sexes of a species.

sibling species: Two closely related, similar-appearing species. Often recognized by ecological and behavioral differences.

species: Taxonomic rank defined as a group of interbreeding individuals reproductively isolated from other such groups (using the "biological species concept").

species complex: Group of closely related, similar-appearing species. In most cases all members are derived from a common ancestor.

stigma: Group of specialized scales, bearing pheromones, located on wings of some male butterflies. Also known as a sex patch.

subapical: Area of the forewing inward from the apex or tip.

submarginal: Relating to the area of the wing inward from the marginal area.

subspecies: Taxonomic rank below the species level, perceived to have consistently different appearance and distribution.

sympatry: Occurring in the same area or range.

tail: Thin projection from the hindwing of many swallowtails and hairstreaks.

tarsal claws: Pair of claws at the tip of the adult butterfly leg (tarsus).

taxon: Unit of organisms concluded to be related and to have characters in common differentiating them from other such units.

taxonomy: Theory and practice of classifying organisms.

thorax: Second or middle body part to which the wings and legs are attached.

tornus: Junction of the inner and outer margins of forewing.

univoltine: Having one flight per year.

valves: Paired structure at the end of the male abdomen designed to hold a female during mating. Also known as **claspers**.

variation: Deviation from the expected or norm, such as differing colors and patterns.

veins: Any of the tubular supports in the butterfly wing.

venter: Undersurface of wings and body.

voltinism: Number of flights per year, usually corresponding to number of generations.

voucher specimen: Specimen retained as a reference for study.

xeric: Habitat that is extremely dry, lacking humidity and water.

Xerothermic: Warm, dry, postglacial interval.

Found below for each museum specimen photographed are: common name; sex; location where captured or, in the case of reared specimens, where the female was caught or egg found; date (the date is given using the standard for lepidopterists: day·month in Roman numerals·year); series of collections in which the specimen existed. When a specimen has been in more than one collection a chronological listing of its location from collection to collection is given, from earliest to latest, with arrows indicating the changes. In the case of reared specimens the following notations are used: the date the egg was found in the wild, denoted "ex egg," the date the egg was laid if a confined female denoted "ex ♀," and the emergence of butterfly from the chrysalis, denoted "em'g'd." When photographs of museum specimens are used outside species accounts, e.g., Special Topics, no data is given.

Abbreviations Used

CMNH Carnegie Museum of Natural History

FEM Frost Entomological Museum of Pennsylvania State University

MGCL McGuire Center for Lepidoptera and Biodiversity of the University of Florida

NMNH United States National Museum of Natural History (Smithsonian)

YPM Yale Peabody Museum of Natural History

30 **Pipevine Swallowtail**—♂—PA, Beaver Co., Beaver, off Ohio River—14·VII·2011—J. L. Monroe → MGCL

31 **Pipevine Swallowtail**—♀—PA, Beaver Co., Beaver, off Ohio River—12·VII·2011—J. L. Monroe → MGCL

32 **Zebra Swallowtail**—♂—WV, Boone Co., Fork Creek WMA, ex egg 30·V·2009—em'g'd 12·III·2010—R. Boscoe → MGCL

33 **Zebra Swallowtail**—♀—WV, Boone Co., Fork Creek WMA, ex egg 30·V·2009—em'g'd 1·III·2010—R. Boscoe → MGCL

34 **Black Swallowtail**—♂—PA, Washington Co., Buffalo Creek—16·VIII·2011—J. L. Monroe → MGCL

35 **Black Swallowtail**—♀—PA, Beaver Co., Industry—4·IX·2010—J. L. Monroe → MGCL

36 **Giant Swallowtail**—♂—PA, Centre Co., Spring Creek—ex egg 9·VI·1981, em'g'd 17·VII·1981—F. D. Fee → MGCL

37 **Giant Swallowtail**—♀—PA, Centre Co., Spring Creek—ex larva 22·VI·1981, em'g'd 28·VII·1981—F. D. Fee→ MGCL

38 **Eastern Tiger Swallowtail**—♂—PA, Beaver Co., Industry—5·VI·2011—J. L. Monroe → MGCL

39 **Eastern Tiger Swallowtail**—♀—PA, Beaver Co., Raccoon Creek State Park—2·VIII·1975—R. W. Surdick → CMNH

40 **Canadian Tiger Swallowtail**—♂—Nova Scotia, Kings Co., Lake Kejimkujik—31·V·1950—D. C. Ferguson → MGCL

41 **Canadian Tiger Swallowtail**—♀—Nova Scotia, Kings Co., Auburn—20·VI·1963—D. C. Ferguson → MGCL

42   **Appalachian Tiger Swallowtail**—♂—PA, Franklin Co., Caledonia—19·V·2001—
      D. M. Wright

43   **Appalachian Tiger Swallowtail**—♀—NC, Clay Co., Buck Creek—14·V·2001—
      R. R. Gatrelle → MGCL

45   **Eastern Tiger Swallowtail** (dark form)—♀—VA, Ritchie Co., Poverty Hollow—
      12·VII·1980—G. Austin → MGCL

45   **Appalachian Tiger Swallowtail** (intermediate form)—♀—WV, Randolph Co., Gladys
      River—18·VI·1984—T. Allen → MGCL

46   **Spicebush Swallowtail**—♂—PA, Beaver Co., Beaver, off Ohio River—13·V·2010—
      J. L. Monroe → MGCL

47   **Spicebush Swallowtail**—♀—PA, Beaver Co., Beaver, off Ohio River—24·VII·2011—
      J. L. Monroe → MGCL

48   **Palamedes Swallowtail**—♂—NC, Pender Co., Holly Shelter Game Lands—29·IV·2012
      —J. L. Monroe → MGCL

49   **Palamedes Swallowtail**—♀—FL, St. John's Co., State Rte. 210—ex egg 1·V·1987,
      em'g'd 28·VI·1987—F. Bodnar → MGCL

52   **Checkered White**—♂—FL, Martin Co., Port Mayaca—ex ♀ 14·V·1982, em'g'd
      10·VI·1982—R. W. Boscoe → MGCL

52   **Checkered White**—♀—IN, Pulaski Co., Winamac Fish and Wildlife Area—7·VII·1977
      —M. Minno

53   **West Virginia White**—♂—PA, Armstrong Co., Spring Church—em'g'd 15·IV·1992—
      F. R. Bodnar → MGCL

53   **West Virginia White**—♀—PA, Armstrong Co., Spring Church—em'g'd 18·IV·1992—
      F. R. Bodnar → MGCL

54   **Cabbage White**—♂—PA, Beaver Co., Industry—26·VI·2011—J. L. Monroe → MGCL

54   **Cabbage White**—♀—PA, Beaver Co., Industry—22·VI·2011—J. L. Monroe → MGCL

55   **Cabbage White** (spring/fall form)—♀—PA, Allegheny Co., Pitts-
      burgh—26·IV·1941—R. W. Surdick → CMNH

56   **Olympia Marble**—♂—PA, Armstrong Co., Spring Church—em'g'd 15·IV·1992—
      F. R. Bodnar → MGCL

56   **Olympia Marble**—♀—PA, Armstrong Co., Spring Church—em'g'd 18·IV·1992—
      F. R. Bodnar → MGCL

57   **Falcate Orangetip**—♂—PA, York Co., G. Pinchot State Park—3·IV·1979—F. D.Fee →
      MGCL

57   **Falcate Orangetip**—♀—PA, Bucks Co., Coffman Hills—2·V·1973—F. D.Fee → MGCL

58   **Clouded Sulphur**—♂—PA, Beaver Co., State Game Lands 173—1·VII·2011—
      J. L. Monroe → MGCL

58   **Clouded Sulphur**—♀—PA, Beaver Co., Industry—9·VII·2010—J. L. Monroe → MGCL

59   **Orange Sulphur**—♂—PA, Beaver Co., Industry—10·VII·2011—J. L. Monroe → MGCL

59   **Orange Sulphur**—♀—PA, Beaver Co., Industry -5·VII·2010—J. L. Monroe → MGCL

60   *alba* **form**—♀—PA, Juniata Co., McCullough's Mills—9·VIII·1994—W. Zanol → CMNH

60   *alba* **form**—♀—PA, Elk Co., 3 km N of Portland Mills—8·VIII·1993—W. Zanol → CMNH

61   **Pink-edged Sulphur**—♂—WV, Tucker Co., 7 mi S of Thomas—11·VII·1983—
      J. E. VanDellen → MGCL

61   **Pink-edged Sulphur**—♀—WV, Pendleton Co., Spring Church—ex ♀ 7·VII·1984,
      em'g'd 3·VI·1985—R. W. Boscoe → MGCL

62   **Southern Dogface**—♂—MO, Jackson Co., near Atherton—27·IX·1965—
      W. A. Hammer → California Academy of Sciences → San Diego Natural History
      Museum→MGCL

62   **Southern Dogface**—♀—MO, Greene Co., Springfield—?·VIII·1968—L. Brown →
      MGCL

63   **Cloudless Sulphur**—♂—FL, Miami-Dade Co., Florida City—ex ♀ 1·X·2003, em'g'd
      28·X·2003—R. W. Boscoe → MGCL

63   **Cloudless Sulphur**—♀—FL, Alachua Co., near Gainesville—15·IX·1973—P. F. Milner
      → MGCL

64   **Orange-barred Sulphur**—♀—FL, Miami-Dade Co., Florida City—ex ♀ 1·X·2003,
      em'g'd 30·X·2003—R. W. Boscoe → MGCL

64   **Orange-barred Sulphur**—♀—FL, Volusia Co., Cassadaga—21·XI·1959—S. V. Fuller →
      MGCL

65   **Sleepy Orange**—♂—WV, Pendleton Co., 4 mi E of Seneca Rocks—20·IX·1984—
      T. Allen → MGCL

65   **Sleepy Orange**—♀—PA, Washington Co., Finleyville—18·VI·1980—MGCL

66   **Little Yellow**—♂—PA, Centre Co., Scotia Barrens—21·VII·1975—F. D. Fee → MGCL

66   **Little Yellow**—♀—PA, Lehigh Co., Bethlehem—5·VIII·1973—F. D. Fee → MGCL

67   **Dainty Sulphur**—♂—TX, Uvalde Co., Uvalde—17·III·1989—F. and J. Preston → MGCL

67   **Dainty Sulphur**—♀—TX, Dallas Co., Dallas—24·VI·1971—I. Finkelstein → MGCL

70   **Harvester**—♂—PA, Bradford Co., LeRoy, Carbon Run—9·VIII·1980—F. D. Fee →
      MGCL

70   **Harvester**—♀—PA, Bradford Co., LeRoy, Carbon Run—22·VII·1980—F. D. Fee →
      MGCL

71   **American Copper**—♂—PA, Indiana Co., 2 mi N of Shelocta—em'g'd 25·VI to
      4·VII·1986—M. G. Douglas → MGCL

71   **American Copper**—♀—PA, Indiana Co., 2 mi N of Shelocta—em'g'd 25·VI to
      4·VII·1986—M. G. Douglas → MGCL

72   **Bronze Copper**—♂—NE, Saunders Co., U of Neb ARDC—M. G. Douglas → MGCL

72   **Bronze Copper**—♀—WI, Oakland Co.—em'g'd 11·VIII to 22·VIII·1977—
      M. G. Douglas → MGCL

73   **Bog Copper**—♂—PA, Carbon Co., Rte. 940 bog—2·VII·1980—D. M. Wright

73   **Bog Copper**—♀—PA, Tioga Co., Reynolds Spring Natural Area—ex ♀ 12·VI·1995,
      em'g'd 28·VI·1996—R. W. Boscoe → MGCL

74   **Great Purple Hairstreak**—♂—FL, Clay Co., Gold Head Branch State Park—ex ♀
      11·III·1981, em'g'd 7·V·1981—F. D. Fee → MGCL

74   **Great Purple Hairstreak**—♀—FL, Liberty Co., Bristol—ex ♀ 15·III·1980, em'g'd
      28·IV·1980—F. D. Fee → MGCL

75   **Coral Hairstreak**—♂—MD, Allegheny Co., Green Ridge State Park—6·VII·1978—
      F. D. Fee → MGCL

75   **Coral Hairstreak**—♀—PA, Centre Co.,—ex ♀ 15·VI·1978, em'g'd 19·VI·1979—
      F. D. Fee → MGCL

76   **Acadian Hairstreak**—♂—PA, Huntingdon Co.—ex ♀ 11·VII·1978, em'g'd 18·VI·1979
      —F. D. Fee → MGCL

76   **Acadian Hairstreak**—♀—PA, Centre Co., Scotia Barrens—8·VII·1980—F. D. Fee →
      MGCL

77   **Edwards' Hairstreak**—♂—PA, Chester Co., Nottingham—25·VI·1973—F. D. Fee →
MGCL

77   **Edwards' Hairstreak**—♀—PA, Centre Co., Scotia Barrens—ex ♀ 17·VI·1978, em'g'd
26·VI·1979—F. D. Fee → MGCL

78   **Banded Hairstreak**—♂—PA, Bucks Co., Ferndale—23·VI·1973—F. D. Fee → MGCL

78   **Banded Hairstreak**—♀—PA, Westmoreland Co., Jones Mills—3·VII·1971—
R. W. Surdick → CMNH

79   **Hickory Hairstreak**—♂—PA, Centre Co., Spring Creek—ex ♀ 17·VI·1978, em'g'd
18·VI·1979—F. D. Fee → MGCL

79   **Hickory Hairstreak**—♀—PA, Centre Co., Spring Creek—11·VII·1974—F. D. Fee →
MGCL

80   **Striped Hairstreak**—♂—PA, Centre Co., State College—ex ♀ 7·VII·1978, em'g'd
29·VI·1979—F. D. Fee → MGCL

80   **Striped Hairstreak**—♀—PA, Centre Co., State College—ex ♀ 7·VII·1978, em'g'd
29·VI·1979—F. D. Fee → MGCL

81   **Oak Hairstreak**—♂—NJ, Ocean Co., Lakehurst—ex ♀ 21·VI·1980, em'g'd
17·VI·1981—R. W. Boscoe → MGCL

81   **Oak Hairstreak**—♀—NJ, Ocean Co., Lakehurst—ex ♀ 21·VI·1980, em'g'd
17·VI·1981—R. W. Boscoe → MGCL

82   **Brown Elfin**—♂—PA, Clearfield Co., 2 mi S of Shawville—ex ♀ 14·V·1987, em'g'd
?·?·1988—F. Bochner → MGCL

82   **Brown Elfin**—♀—PA, Clearfield Co., 2 mi S of Shawville—ex ♀ 14·V·1987, em'g'd
?·?·1988—F. Bochner → MGCL

83   **Hoary Elfin**—♂—NJ, Burlington Co., Lebanon State Forest—13·V·1972—M. Douglas
→ MGCL

83   **Hoary Elfin**—♀—NJ, Burlington Co., Lebanon State Forest—13·V·1972—M. Douglas
→ MGCL

84   **Frosted Elfin**—♂—PA, Centre Co., Scotia Barrens—ex ♀ 14·V·1979, em'g'd
29·III·1980—F. D. Fee → MGCL

84   **Frosted Elfin**—♀—PA, Centre Co., Scotia Barrens—ex ♀ 14·V·1979, em'g'd
15·IV·1980—F. D. Fee → MGCL

85   **Henry's Elfin**—♂—PA, York Co., Gifford Pinchot State Park—ex ♀ 30·IV·1979, em'g'd
1·IV·1980—MGCL

85   **Henry's Elfin**—♀—PA, York Co., Gifford Pinchot State Park—ex ♀ 30·IV·1979, em'g'd
26·II·1980—MGCL

86   **Eastern Pine Elfin**—♂—NJ, Ocean Co., Lakehurst—ex ♀ 4·V·1985, em'g'd
10·IV·1986—MGCL

86   **Eastern Pine Elfin**—♀—NJ, Ocean Co., White Plains, Rte. 72—6·V·1972—F. D. Fee →
MGCL

87   **Juniper Hairstreak**—♂—PA, Montgomery Co., Sumneytown—ex ♀ 1·VI·1973, em'g'd
9·VII·1973—F. D. Fee → MGCL

87   **Juniper Hairstreak**—♀—PA, Montgomery Co., Sumneytown—ex ♀ 1·VI·1973, em'g'd
15·VII·1973—F. D. Fee → MGCL

88   **White M Hairstreak**—♂—PA, Centre Co., Black Moshannon State Park—ex ♀
22–31·VIII·2006, em'g'd 25·X·2006—F. D. Fee

88 **White M Hairstreak**—♀—PA, Centre Co., Black Moshannon State Park—ex ♀ 22–31·VIII·2006, em'g'd 25·X·2006—F. D. Fee

89 **Gray Hairstreak**—♂—PA, Centre Co., Scotia Barrens—ex ♀ 14·VII·1980, em'g'd 20·VIII·1980—F. D. Fee → MGCL

89 **Gray Hairstreak**—♀—PA, Centre Co., Scotia Barrens—ex ♀ 14·VII·1980, em'g'd 22·VIII·1980—F. D. Fee → MGCL

90 **Red-banded Hairstreak**—♂—PA, Montgomery Co., Flourtown—ex ♀ 20·VIII·1992, em'g'd 27·X·1992—R. W. Boscoe → MGCL

90 **Red-banded Hairstreak**—♀—PA, Montgomery Co., Flourtown—ex ♀ 20·VIII·1992, em'g'd 24·X·1992—R. W. Boscoe → MGCL

91 **Early Hairstreak**—♂—Canada, NB, York Co., Prince William - em'g'd 20·IV·1992— R. P. Webster → MGCL

91 **Early Hairstreak**—♀—Canada, NB, York Co., Prince William - em'g'd 25·VII·1991— R. P. Webster → MGCL

92 **Marine Blue**—♂—AZ, Santa Rites, Madera Canyon—7·IX·1970—M. Douglas → J. D. Turner → MGCL

92 **Marine Blue**—♀—Mexico, Oaxaca, Riso de Oro, Rte. 190—7·VI·1994—M. Douglas → J. D. Turner → MGCL

93 **Eastern Tailed-Blue**—♂—PA, Allegheny Co., Bethel Park—28·VII·1951— R. W. Surdick → CMNH

93 **Eastern Tailed-Blue**—♀—PA, Allegheny Co., Bethel Park—28·VII·1951— R. W. Surdick → CMNH

94 **Spring Azure**—♂—PA, Bucks Co., State Game Lands 196—15·IV·1995—D. M. Wright

94 **Spring Azure**—♀—PA, Berks Co., Seisholtzville, Doe Mtn.—1st instar 7·V·2006, fridge 2 mos., adult 12·XI·2006—D. M. Wright

95 **Northern Spring Azure**—♂—PA, Berks Co., Blue Mtn.—14·IV·2000—D. M. Wright

95 **Northern Spring Azure**—♀—PA, Monroe Co., Pocono Pines—ex ♀ 8·V·1993, em'g'd 15·IV·1994—D. M. Wright

96 **Summer Azure**—♂—PA, Monroe Co., State Game Lands 127—1·VIII·1998— D. M. Wright

96 **Summer Azure**—♀—PA, Bucks Co., Ferndale—ex larva 20·VI·1995, em'g'd 15·VII·1995—R. W. Boscoe → MGCL

97 **Cherry Gall Azure**—♂—PA, Monroe Co., State Game Lands 127—24·V·1997— D. M. Wright

97 **Cherry Gall Azure**—♀—PA, Monroe Co., State Game Lands 127—26·V·2000— D. M. Wright

98 **Appalachian Azure**—♂—PA, Bucks Co., State Game Landss #157—1·VI·1990— D. M. Wright

98 **Appalachian Azure**—♀—WV, Pendleton Co., Spruce Knob, Rte. 103—27·V·2007— D. M. Wright

99 **Dusky Azure**—♂—WV, Boone Co., Fork Creek Public Hunting Area— ex larva 20·IV·1990, pupa 8·V·1990, fridge 4 mos., adult 7·X·1990—D. M. Wright

99 **Dusky Azure**—♀—WV, Boone Co., Fork Creek Public Hunting Area— ex larva 20·IV·1990, pupa 8·V·1990, fridge 4 mos., adult 9·X·1990—D. M. Wright

102 **Silvery Blue**—♂—PA, Centre Co., Benner Twp, Spring Creek—ex ♀ 26·IV·1987, em'g'd 30·I·1988—R. W. Boscoe → MGCL

102 **Silvery Blue**—♀—PA, Centre Co., Scotia Barrens—ex ♀ 26·IV·1987, em'g'd
30·I·1988—R. W. Boscoe → MGCL

103 **Karner Blue**—♂—NY, Albany Co., Albany—ex ♀ 18·VII·1987, em'g'd 29·V·1988—
R. W. Boscoe → MGCL

103 **Karner Blue**—♀—NY, Albany Co., Albany—ex ♀ 18·VII·1987, em'g'd 29·V·1988—
R. W. Boscoe → MGCL

106 **Northern Metalmark**—♂—PA, Centre Co., "The Rock"—11·V·1980—F. D. Fee →
MGCL

106 **Northern Metalmark**—♀—PA, Centre Co., Benner Twp, Spring Creek—ex ♀
14·VII·1984—R. W. Boscoe → MGCL

107 **Swamp Metalmark**—♂—MI, Willis—9·VII·1933—Paratype #23—W. S. McAlpine →
CMNH

107 **Swamp Metalmark**—♀—MI, Oakland Co., Mahopec—24·VII·1932—Paratype
#32—W. S. McAlpine → CMNH

111 **American Snout**—♂—PA, Centre Co., Benner Twp, Spring Creek—15·VIII·1983—
F. D. Fee → MGCL

111 **American Snout**—♀—PA, Montgomery Co., Flourtown—ex ♀ 2·VII·1984, em'g'd
26·VII·1984—R. W. Boscoe → MGCL

112 **Monarch**—♂—PA, Beaver Co., Wolf Run Rd.—9·VIII·2011—J. L. Monroe → MGCL

113 **Monarch**—♀—PA, Monroe Co., State Game Lands 127—em'g'd 20·IX·2011—
D. M. Wright

114 **Queen**—♂—FL, Miami-Dade Co., Homestead—28·IV·1975—L. L. Harris → MGCL

115 **Queen**—♀—FL, Monroe Co., Hwy 94, near Pinecrest—22·III·1979—R. F. Denno →
MGCL

116 **Gulf Fritillary**—♂—FL, Monroe Co., North Key Largo—ex ♀ 9·XI·1982, em'g'd
17·XII·1982—R. W. Boscoe → MGCL

116 **Gulf Fritillary**—♀—FL, Monroe Co., North Key Largo—ex ♀ 9·XI·1982, em'g'd
17·XII·1982—R. W. Boscoe → MGCL

117 **Variegated Fritillary**—♂—PA, Fayette Co.—31·VIII·1980—R. W. Surdick → CMNH

117 **Variegated Fritillary**—♀—NC, Pender Co., Holly Shelter Game Lands—27·IV·2012—
J. L. Monroe → MGCL

118 **Great Spangled Fritillary**—♂—PA, Beaver Co., State Game Lands 285—2·VI·2010—
J. L. Monroe → MGCL

119 **Great Spangled Fritillary**—♀—PA, Beaver Co., State Game Lands 285—22·VI·2011—
J. L. Monroe → MGCL

120 **Aphrodite Fritillary**—♂—PA, Centre Co., Black Moshannon State Park—ex ♀
17·VII·1982, em'g'd 11·IV·1983—R. W. Boscoe → MGCL

121 **Aphrodite Fritillary**—♀—PA, Centre Co., Colyer Lake—30·VI·2077—MGCL

122 **Atlantis Fritillary**—♂—PA, Forest Co., Marienville, Buzzard Swamp—29·VI·2011—
J. L. Monroe → MGCL

123 **Atlantis Fritillary**—♀—PA, Clarion Co., Cook Forest—2·VIII·1998—R. W. Surdick →
CMNH

124 **Regal Fritillary**—♂—PA, Dauphin Co., Linglestown—30·VI·1918—G. S. Anderson →
FEM

125 **Regal Fritillary**—♀—PA, Allegheny Co.—18·VII·1947—R. W. Surdick → CMNH

126 **Diana Fritillary**—♂—VA, Montgomery Co., Poverty Hollow—ex ♀ 19·VIII·1978, em'g'd 11·VI·1979—F. D. Fee → MGCL

127 **Diana Fritillary**—♀—VA, Montgomery Co., Poverty Hollow—ex ♀ 19·VIII·1978, em'g'd 20·IV·1979—F. D. Fee → MGCL

128 **Silver-bordered Fritillary**—♂—PA, Centre Co., Colyer Lake—ex ♀ 17·VII·1982, em'g'd 17·VIII·1982—R. W. Boscoe → MGCL

128 **Silver-bordered Fritillary**—♀—PA, Centre Co., Colyer Lake—ex ♀ 17·VII·1982, em'g'd 17·VIII·1982—R. W. Boscoe → MGCL

129 **Meadow Fritillary**—♂—PA, Beaver Co., State Game Lands 173—18·VIII·2011— J. L. Monroe → MGCL

129 **Meadow Fritillary**—♀—PA, Tioga Co., Debnar Twp, Baldwin Run—6·VIII·1968— G. F. Patterson → FEM

130 **Gorgone Checkerspot**—♂—SC, North Orangeburg Co.,—ex ♀ 28·IV·1995, em'g'd 22·V·1995—R. W. Boscoe → MGCL

130 **Gorgone Checkerspot**—♀—SC, North Orangeburg Co.,—ex ♀ 28·IV·1995, em'g'd 4·VI·1995—R. W. Boscoe → MGCL

131 **Silvery Checkerspot**—♂—MD, Cecil Co., Conowingo—9·VIII·1973—F. D. Fee → MGCL

131 **Silvery Checkerspot**—♀—PA, Centre Co., Benner Twp, Spring Creek—19·VI·1973— F. D. Fee → MGCL

132 **Harris' Checkerspot**—♂—PA, Clinton Co., Tamarack—14·VI·1975—F. D. Fee → MGCL

132 **Harris' Checkerspot**—♀—PA, Centre Co., Black Moshannon State Park—ex larva 1·VI·1978, em'g'd 13·VI·1978—F. D. Fee → MGCL

133 **Baltimore Checkerspot**—♂—PA, Tioga Co., Baldwin Run—e'mg'd 7·VII·1971— G. F. Patterson → FEM

133 **Baltimore Checkerspot**—♀—PA, Somerset Co., Kooser Park—reared from larva 27·VI·1952—R. W. Surdick → CMNH

134 **Pearl Crescent** (summer form)—♂—NJ, Bergen Co., Bergen Swamp—22·VII·1974— R. W. Surdick → CMNH

134 **Pearl Crescent** (summer form)—♀—PA, Tioga Co., Covington Twp—15·VI·1958— G. F. Patterson → FEM

134 **Pearl Crescent** (fall/spring form)—♂—NC, Pender Co., Holly Shelter Game Land— 26·IV·2012—J. L. Monroe → MGCL

134 **Pearl Crescent** (fall/spring form)—♀—PA, Beaver Co., Raccoon Creek State Park— 22·V·1977—R. W. Surdick → CMNH

136 **Northern Crescent**—♂—PA, Tioga Co., Middlebury Twp, Stephenhouse Run— 23·VI·1967—G. F. Patterson → FEM

136 **Northern Crescent**—♀—PA, Tioga Co., Delmar Twp, Baldwin Run—8·VI·1968— G. F. Patterson → FEM

137 **Tawny Crescent**—♂—PA, Scranton—30·V·1906—M. Rothke → Marloff → YPM

137 **Tawny Crescent**—♀—PA, Scranton—18·VI·1902—H. P. Wilhelm → YPM

138 **Common Buckeye**—♂—IN, Wayne Co., Richmond—20·VIII·2011—D. M. Wright

138 **Common Buckeye**—♀—IN, Wayne Co., Richmond—20·VIII·2011—D. M. Wright

139 **Common Buckeye** (*rosa* form)—♂—AR, Grant Co., Sheridan—8·X·1975— M. C. Douglas → MGCL

139   **Common Buckeye** (*rosa* form)—♀—PA, Beaver Co., off Ohio River, Industry—
23·X·2012—J. L. Monroe → MGCL

140   **Question Mark** (summer form)—♂—PA, Allegheny Co., Oak Station—4·VII·1919—
Marloff → CMNH

140   **Question Mark** (summer form)—♀—PA, Beaver Co., off Ohio River, Industry—
1·VIII·2011—J. L. Monroe → MGCL

141   **Question Mark** (fall/spring form)—♂—PA, Beaver Co., off Ohio River, Industry—
12·IX·2010—J. L. Monroe → MGCL

141   **Question Mark** (fall/spring form)—♀—PA, Centre Co., Benner Twp, Spring Creek—
12·IX·1977—R. W. Boscoe → MGCL

142   **Eastern Comma** (summer form)—♂—PA, Beaver Co., off Ohio River, Industry—
18·VI·2011—J. L. Monroe → MGCL

142   **Eastern Comma** (summer form)—♀—PA, Centre Co., Benner Twp, Spring Creek—
ex egg 8·V·1979, em'g'd 12·VI·1979—F. D. Fee → MGCL

143   **Eastern Comma** (fall/spring form)—♂—PA, Centre Co., Benner Twp, Spring Creek—
23·IX·1977—F. D. Fee → MGCL

143   **Eastern Comma** (fall/spring form)—♀—PA, Beaver Co., Raccoon Creek State Park—
13·IX·1994—R. W. Surdick → CMNH

144   **Green Comma**—♂—PA, Monroe Co., Tobyhanna—22·VII·1929—E. Shoemaker →
NMNH

144   **Green Comma**—♀—PA, Monroe Co., Tobyhanna—25·VII·1929—E. Shoemaker →
NMNH

145   **Gray Comma**—♂—PA, Huntingdon Co., Laural Run Rd.,—ex ♀ 25·VII·1994, em'g'd
17·IX·1994—R. W. Boscoe → MGCL

145   **Gray Comma**—♀—PA, Huntingdon Co., Laural Run Rd.,—ex ♀ 25·VII·1994, em'g'd
17·IX·1994—R. W. Boscoe → MGCL

148   **Compton Tortoiseshell**—♂—PA, Centre Co., Waddle—ex egg 30·IV·1993, em'g'd
4·VI·1993—R. W. Boscoe → MGCL

148   **Compton Tortoiseshell**—♀—PA, Centre Co., Waddle—ex egg 30·IV·1993, em'g'd
5·VI·1993—R. W. Boscoe → MGCL

149   **California Tortoiseshell**—♂—CA, Napa Co., Oakville—em'g'd 23·IV·1971—
R. F. Denno and D. Chandler → MGCL

149   **California Tortoiseshell**—♀—CA, Napa Co., Oakville—em'g'd 23·IV·1971—
R. F. Denno and D. Chandler → MGCL

150   **Mourning Cloak**—♂—PA, Centre Co., Benner Twp, Spring Creek—30·VIII·1976—
F. D. Fee → MGCL

150   **Mourning Cloak**—♀—NV, Clark Co., Las Vegas—em'g'd 12·IV·1986—G. T. Austin →
MGCL

151   **Milbert's Tortoiseshell**—♂—Canada, New Brunswick—22·VII·1960—MGCL

151   **Milbert's Tortoiseshell**—♀—PA, Centre Co., Benner Twp, Spring Creek—ex egg
1·V·1977, em'g'd 20·V·1977—R. W. Boscoe → MGCL

152   **American Lady**—♂—NY, Suffolk Co., Smithtown—M. Leeburger → MGCL

152   **American Lady**—♀—?·IX·1952—MGCL

153   **Painted Lady**—♂—CA, Mono Co.—22·VII·1970—R. F. Denno → MGCL

153   **Painted Lady**—♀—PA, Centre Co., Benner Twp, Spring Creek—ex egg 5·VII·1983,
em'g'd 19·VII·1983—F. D. Fee → MGCL

154 **Red Admiral**—♂—PA, Centre Co., Benner Twp, Spring Creek—em'g'd 20·VII·1979—
F. D. Fee → MGCL

154 **Red Admiral**—♀—PA, Centre Co., Benner Twp, Spring Creek—5·VII·1979—F. D. Fee
→ MGCL

156 **White Admiral**—♂—PA, Bradford Co., 2 mi S of LeRoy Twp, Carbon Run Bog—
9·VIII·1980—F. D. Fee → MGCL

156 **White Admiral**—♀—PA, Tioga Co., Arnot—10·VII·1993—R. W. Boscoe→ MGCL

157 **Red-spotted Purple**—♂—PA, Beaver Co., Wolf Run Rd.—6·VII·2010—J. L. Monroe →
MGCL

157 **Red-spotted Purple**—♀—PA, Lycoming Co., Cogan House Twp—27·VI·1987—
G. F. Patterson → MGCL

158 **Red-spotted Purple** (intergrade)—PA, Tioga Co., Morris Twp, Rattlesnake Rd.—
21·VI·1996—R. Hirzel → MGCL

158 **Red-spotted Purple** (intergrade)—PA, Lawrence Co., Slippery Rock Creek—
15·VIII·1939—J. Bauer → MGCL

158 **Common Wood Nymph** (intergrade)—PA, Tioga Co., Covington Twp, Elkin Run
Rd.—25·VII·1967—G. F. Patterson → FEM

158 **Common Wood Nymph** (intergrade)—PA, Somerset Co., Forbes State Forest—20·VII·1983—CMNH

159 **Viceroy**—♂—PA, Beaver Co., Wolf Run Rd.—6·VII·2010—J. L. Monroe → MGCL

159 **Viceroy**—♀—PA, Washington Co.—11·VIII·1951—R. W. Surdick → CMNH

160 **Hackberry Emperor**—♂—PA, Montgomery Co., Flourtown—ex larva 10·V·1976,
em'g'd 1·VI·1976—R. W. Boscoe → MGCL

160 **Hackberry Emperor**—♀—PA, Centre Co., Benner Twp, Spring Creek—ex ♀ 7·V·1978,
em'g'd 13·VIII·1978—R. W. Boscoe → MGCL

161 **Tawny Emperor**—♂—PA, Centre Co., Benner Twp, Spring Creek—em'g'd
27·IX·1978—F. D. Fee → MGCL

161 **Tawny Emperor**—♀—PA, Centre Co., Benner Twp, Spring Creek—ex ♀ 18·VII·1978,
em'g'd 2·IX·1978—F. D. Fee → MGCL

162 **Tawny Emperor**, form *proserpina*—♀—PA, Washington Co., Lawrence—3·VII·1950—
R. W. Surdick → CMNH

163 **Northern Pearly-Eye**—♂—PA, Forest Co., Buzzard Swamp, Marienville—29·VI·2011
—J. L. Monroe → MGCL

163 **Northern Pearly-Eye**—♀—PA, Tioga Co., Covington Twp, Elkin Run Rd.—
22·VII·1967—G. F. Patterson → FEM

164 **Eyed Brown**—♂—PA, Tioga Co., Gaines Twp, Upper Lick Run—22·VII·1972—G. F.
Patterson → FEM

164 **Eyed Brown**—♀—PA, Centre Co., Colyer, Potter Twp—9·VII·1974—F. D. Fee →
MGCL

165 **Appalachian Brown**—♂—PA, Lehigh Co., Bethlehem—10·VII·1973—F. D. Fee →
MGCL

165 **Appalachian Brown**—♀—PA, Monroe Co., State Game Lands 127—6·VII·1989—
D. M. Wright

166 **Carolina Satyr**—♂—KY, Letcher Co., Whiteburg—25·VII·1971—R. W. Surdick →
CMNH

166 **Carolina Satyr**—♀—FL, Collier Co., Naples—17·III·1976—G. F. Patterson → MGCL

167 **Little Wood Satyr**—♂—PA, Tioga Co., Covington Twp Mead Gun Run—25·V·1977
—G. F. Patterson → FEM

167 **Little Wood Satyr**—♀—PA, Tioga Co., Delmar Twp, Canada Run—14·VI·1963—
G. F. Patterson → FEM

168 **Common Ringlet**—♂—PA, Pike Co., Rte. 6, 1 mi W of Rte. 434—14·VIII·1995—
D. M. Wright

168 **Common Ringlet**—♀—PA, Monroe Co., State Game Lands 127—29·V·2004—
D. M. Wright

169 **Common Wood Nymph**—♂—PA, Monroe Co., Gilbert—2·VIII·1972—F. D. Fee →
FEM

169 **Common Wood Nymph**—♀—PA, Beaver Co., Raccoon Creek State Park—
16·VII·1983—R. W. Surdick → CMNH

172 **Silver-spotted Skipper**—♂—PA, Montgomery Co., Flourtown—ex larva 27·VIII·2002,
em'g'd 20·II·2003—R. W. Boscoe → MGCL

172 **Silver-spotted Skipper**—♀—PA, Centre Co., Scotia Barrens—ex larva 22·IV·1978,
em'g'd 28·V·1978—R. W. Boscoe → MGCL

173 **Long-tailed Skipper**—♂—FL, Miami-Dade Co., Homestead, Camp Owassa-Bauer—
24·XII·1979—R. W. Boscoe → MGCL

173 **Long-tailed Skipper**—♀—SC, Beaufort Co., Burton—31·X·1989—D. L. Bauer →
J. D. Turner → MGCL

174 **Golden-banded Skipper**—♂—WV, Boone Co., Blue Wilderness Fork and Joe's Branch
Fork, Nellis—4·VI·1985—T. Allen → MGCL

174 **Golden-banded Skipper**—♀—GA, Grady Co., Sherwood Plantation—11·III·1965—
CMNH

175 **Hoary Edge**—♂—PA, Montgomery Co., Fort Washington State Park, Fort Hill—
12·VI·1976—R. W. Boscoe → MGCL

175 **Hoary Edge**—♀—PA, Montgomery Co., Green Lane—23·VI·1973—F. D. Fee → MGCL

176 **Northern Cloudywing**—♂—PA, Huntingdon Co., Martin Gap Rd.—ex ♀ 29·V·2004,
em'g'd 10·VIII·2004—R. W. Boscoe → MGCL

176 **Northern Cloudywing**—♀—PA, Huntingdon Co., Martin Gap Rd.—ex ♀ 29·V·2004,
em'g'd 11·VIII·2004—R. W. Boscoe → MGCL

177 **Northern Cloudywing**—♂—PA, Greene Co., Waynesburg—1·VI·1968—R. W. Surdick → CMNH

177 **Northern Cloudywing**—♂—PA, Allegheny Co., Pittsburgh—9·VI·1942—
R. W. Surdick → CMNH

178 **Southern Cloudywing**—♂—NC, Moore Co., Foxfire Village—ex ♀ 29·IV·1995, em'g'd
28·VI·1995—R. W. Boscoe → MGCL

178 **Southern Cloudywing**—♀—NC, Moore Co., Foxfire Village—ex ♀ 29·IV·1995, em'g'd
3·VII·1995—R. W. Boscoe → MGCL

179 **Confused Cloudywing**—♂—NC, Moore Co., Foxfire Village—ex ♀ 29·IV·1995,
em'g'd 29·VI·1995—R. W. Boscoe → MGCL

179 **Confused Cloudywing**—♀—NC, Moore Co., Foxfire Village—ex ♀ 29·IV·1995, em'g'd
29·VI·1995—R. W. Boscoe → MGCL

182 **Dreamy Duskywing**—♂—PA, Centre Co., Scotia Barrens—ex ♀ 2·VI·1990, em'g'd
2·V·1991—R. W. Boscoe → MGCL

Locations for Photographed Specimens

182   **Dreamy Duskywing**—♀—PA, Centre Co., Scotia Barrens—ex ♀ 2·VI·1990, em'g'd
       2·V·1991—R. W. Boscoe → MGCL

183   **Sleepy Duskywing**—♂—PA, Huntingdon Co., McAlvey's Fort—ex ♀ 11·V·2001,
       em'g'd 27·VIII·2001—R. W. Boscoe → MGCL

183   **Sleepy Duskywing**—♀—PA, Huntingdon Co., Miller Twp, Martin Gap Road—ex ♀
       30·IV·2003, em'g'd 5·VIII·2003—R. W. Boscoe → MGCL

184   **Juvenal's Duskywing**—♂—PA, Greene Co.—27·IV·1957—R. W. Surdick → CMNH

184   **Juvenal's Duskywing**—♀—PA, Beaver Co., Raccoon Creek State Park—13·V·1993—
       R. W. Surdick → CMNH

185   **Horace's Duskywing**—♂—PA, Centre Co., Scotia Barrens—1·VIII·1978—F. D. Fee →
       MGCL

185   **Horace's Duskywing**—♀—MD, Allegheny Co., Green Ridge State Forest—3·V·1978—
       F. D. Fee → MGCL

186   **Mottled Duskywing**—♂—PA, Chester Co., Nottingham—10·V·1973—MGCL

186   **Mottled Duskywing**—♀—NC, Moore Co., Foxfire Village—ex ♀ 12·IV·1998, em'g'd
       23·VI·1998—R. W. Boscoe → MGCL

187   **Zarucco Duskywing**—♂—FL, Manatee Co., County Line Rd.—8·IX·1980—
       S. R. Steinhauser and L. M. Steinhauser → MGCL

187   **Zarucco Duskywing**—♀—FL, Manatee Co., County Line Rd.—12·IX·1979—
       S. R. Steinhauser and L. M. Steinhauser → MGCL

188   **Funereal Duskywing**—♂—AZ, Santa Cruz Co., Upper Madera Canyon, Santa Rita
       Mtns.—19·IX·1977—D. L. Lindsley → MGCL

188   **Funereal Duskywing**—♀—CA, San Diego Co., Mt. Palomar—26·VI·1987—
       D. L. Lindsley → MGCL

189   **Columbine Duskywing**—♂—PA, Greene Co., Waynesburg—1·VI·1968—
       R. W. Surdick → CMNH

189   **Columbine Duskywing**—♀—PA, Allegheny Co., Pittsburgh—9·VI·1942—
       R. W. Surdick → CMNH

190   **Wild Indigo Duskywing**—♂—NC, Moore Co., Foxfire Village—ex ♀ 29·IV·1995,
       em'g'd 29·VI·1995—R. W. Boscoe → MGCL

190   **Wild Indigo Duskywing**—♀—NC, Moore Co., Foxfire Village—ex ♀ 29·IV·1995,
       em'g'd 29·VI·1995—R. W. Boscoe → MGCL

191   **Persius Duskywing**—♂—PA, Centre Co., Benner Twp, Spring Creek—ex larva
       20·VI·1979, em'g'd 28·VII·1979—R. W. Boscoe → MGCL

191   **Persius Duskywing**—♀—PA, Centre Co., Benner Twp, Spring Creek—ex larva
       17·VIII·1978, em'g'd 16·IX·1978—R. W. Boscoe → MGCL

194   **Hayhurst's Scallopwing**—♂—VA, Suffolk—ex ♀ 7·VIII·1991, em'g'd 6·X·1991—
       R. W. Boscoe → MGCL

194   **Hayhurst's Scallopwing**—♀—VA, Suffolk—ex ♀ 7·VIII·1991, em'g'd 7·X·1991—
       R. W. Boscoe → MGCL

195   **Appalachian Grizzled Skipper**—♂—VA, Montgomery Co., Poverty Hollow—
       13·IV·1978—F. D. Fee → MGCL

195   **Appalachian Grizzled Skipper**—♀—VA, Montgomery Co., Poverty Hollow—
       13·IV·1978—F. D. Fee → MGCL

196   **Common Checkered-Skipper**—♂—PA, Centre Co., Spring Creek Benner Twp—
       2·IX·1983—F. D. Fee → MGCL

196   **Common Checkered-Skipper**—♀—PA, Montgomery Co., Flourtown—ex egg 24·VII·1988, em'g'd 29·VIII·1988—R. W. Boscoe → MGCL

197   **Common Sootywing**—♂—PA, Allegheny Co., Pittsburgh—27·VII·1947— R. W. Surdick → MGCL

197   **Common Sootywing**—♀—PA, Philadelphia Co., Clark Park—ex ♀ 2·V·1988, em'g'd 9·VII·1988—R. W. Boscoe → MGCL

198   **Arctic Skipper**—♂—Canada, New Brunswick, Edmundston—19·VI·1961—H. Hensel → MGCL

198   **Arctic Skipper**—♀— Canada, New Brunswick, Trout River—30·VI·1961—H. Hensel → MGCL

200   **Swarthy Skipper**—♂—PA, Montgomery Co., Flourtown—ex ♀ 11·VI·1989, em'g'd 29·VII·1989—R. W. Boscoe → MGCL

200   **Swarthy Skipper**—♀—PA, Greene Co.—3·VIII·1957—R. W. Surdick → CMNH

201   **Clouded Skipper**—♂—VA, Suffolk—ex ♀ 23·VIII·1988, em'g'd 2·X·1988— R. W. Boscoe → MGCL

201   **Clouded Skipper**—♀—VA, Suffolk—ex ♀ 23·VIII·1988, em'g'd 4·X·1988— R. W. Boscoe → MGCL

202   **Least Skipper**—♂—PA, Montgomery Co., Flourtown—ex ♀ 24·IX·1987, em'g'd 6·X·1987—R. W. Boscoe → MGCL

202   **Least Skipper**—♀—PA, Montgomery Co., Flourtown—ex ♀ 24·IX·1987, em'g'd 27·X·1987—R. W. Boscoe → MGCL

203   **European Skipper**—♂—PA, Monroe Co., Pocono Pines—19·VII·1992—D. M. Wright

203   **European Skipper**—♀—PA, Centre Co., State College—26·VI·1978—F. D. Fee → MGCL

204   **Fiery Skipper**—♂—FL, Monroe Co., North Key Largo—ex ♀ 31·X·1989, em'g'd 18·XII·1989—R. W. Boscoe → MGCL

204   **Fiery Skipper**—♀—FL, Miami-Dade Co., Owassa-Bauer Camp—31·XII·1979— F. D. Fee → MGCL

205   **Fiery Skipper**—♀—PA, Westmoreland Co., 1 mi N of Jones Mills—23·VIII·1995— R. W. Surdick → CMNH

205   **Fiery Skipper**—♀—NC, Buxton Co., Cape Hatteras—29·VIII·1975—R. W. Surdick → CMNH

206   **Sachem**—♂—TX, Dallas Co., vic. Irving—16·IV·2011—R. A. Rahn → CMNH

206   **Sachem**—♀—TX, Dallas Co., vic. Irving—16·IV·2011—R. A. Rahn → CMNH

207   **Sachem**—♂—PA, Montgomery Co., Flourtown—ex ♀ 21·VIII·2200, em'g'd 8·X·2200—R. W. Boscoe → MGCL

207   **Sachem**—♀—MD, Charles Co., Doucaster State Forest—24·VII·1974—R. W. Surdick → CMNH

208   **Leonard's Skipper**—♂—PA, Huntingdon Co., Barree Twp, Ridge and Nursery Rds.—17·VIII·1978—F. D. Fee → MGCL

208   **Leonard's Skipper**—♀—PA, Montgomery Co., Whipple Dam State Park—ex ♀ 26·VIII·1989, em'g'd 8·II·1990—R. W. Boscoe → MGCL

209   **Cobweb Skipper**—♂—PA, Chester Co., Nottingham—10·V·1973—F. D. Fee → MGCL

209   **Cobweb Skipper**—♀—PA, Montgomery Co., Sumneytown—12·V·1973—F. D. Fee → MGCL

210   **Dotted Skipper**—♂—FL, Levy Co., 8 mi SW of Williston—ex ♀ 8·X·1998, em'g'd 22·XII·1998—R. W. Boscoe → MGCL

210 **Dotted Skipper**—♀—FL, Levy Co., 8 mi SW of Williston—ex ♀ 8·X·1998, em'g'd
25·I·1999—R. W. Boscoe → MGCL

212 **Indian Skipper**—♂—PA, Beaver Co., Raccoon Creek State Park—31·V·1981—
R. W. Surdick → CMNH

212 **Indian Skipper**—♀—PA, Beaver Co., Raccoon Creek State Park—27·V·2002—
R. W. Surdick → CMNH

213 **Indian Skipper**—♂—PA, Centre Co., Scotia Barrens—4·VI·1978—F. D. Fee → MGCL

213 **Indian Skipper**—♀—PA, Centre Co., Colyer Lake—28·V·1975—F. D. Fee → MGCL

214 **Peck's Skipper**—♂—PA, Montgomery Co., Flourtown—ex ♀ 10·VIII·1987, em'g'd
26·X·1987—R. W. Boscoe → MGCL

214 **Peck's Skipper**—♀—PA, Montgomery Co., Flourtown—ex ♀ 10·VIII·1987, em'g'd
8·X·1987—R. W. Boscoe → MGCL

215 **Peck's Skipper**—♂—WV, Preston Co., Pine Swamp—R. W. Surdick → CMNH

215 **Peck's Skipper**—♀—PA, Tioga Co., Stephenhouse Run, Middlebury Twp—
4·VII·1979—G. Patterson → FEM

216 **Tawny-edged Skipper**—♂— PA, Greene Co., Mapletown—30·V·1957—R. W. Surdick
→ CMNH

216 **Tawny-edged Skipper**—♀—PA, Washington Co. —18·VIII·1951 —R. W. Surdick →
CMNH

217 **Crossline Skipper**—♂—PA, Allegheny Co., —6·VI·1948—R. W. Surdick → CMNH

217 **Crossline Skipper**—♀—PA, Beaver Co., Raccoon Creek SP—2·VII·1954—
R. W. Surdick → CMNH

218 **Tawny-edged Skipper**—♂—PA, Montgomery Co., Flourtown—27·V·1975—
R. W. Boscoe → MGCL

218 **Tawny-edged Skipper**—♀—PA, Montgomery Co., Flourtown—ex egg 10·VIII·1987,
em'g'd 30·X·1987—R. W. Boscoe → MGCL

218 **Crossline Skipper**—♂—NJ, Ocean Co., Lakehurst—ex ♀ 26·VI·1988, em'g'd
28·IX·1988—R. W. Boscoe → MGCL

218 **Crossline Skipper**— ♀—NJ, Ocean Co., Lakehurst—ex ♀ 26·VI·1988, em'g'd
28·IX·1988—R. W. Boscoe → MGCL

219 **Long Dash**—♂—PA, Monroe Co., Warnertown—4·VII·1993—D. M. Wright

219 **Long Dash**—♀—PA, Monroe Co., State Game Lands 127—27·VI·1998—D. M. Wright

220 **Whirlabout**—♂—SC, Berkeley Co., Monk's Corner—ex ♀ 4·VI·1996, em'g'd
20·VII·1996—R. W. Boscoe → MGCL

220 **Whirlabout**—♀—SC, Berkeley Co., Monk's Corner—ex ♀ 4·VI·1996, em'g'd
20·VII·1996—R. W. Boscoe → MGCL

221 **Little Glassywing**—♂—PA, Montgomery Co., Flourtown—ex ♀ 26·VI·1989, em'g'd
26·IX·1989—R. W. Boscoe → MGCL

221 **Little Glassywing**—♀—PA, Montgomery Co., Flourtown—ex ♀ 26·VI·1989, em'g'd
24·IX·1989—R. W. Boscoe → MGCL

222 **Northern Broken-Dash**—♂—PA, Montgomery Co., Flourtown—ex ♀ 21·VI·1998,
em'g'd 30·IX·1998—R. W. Boscoe → MGCL

222 **Northern Broken-Dash**—♀—PA, Montgomery Co., Flourtown—ex ♀ 28·VI·1988,
em'g'd 25·IX·1988—R. W. Boscoe → MGCL

223 **Southern Broken-Dash**—♂—FL, Monroe Co., Key Largo—23·V·1985—R. W. Surdick
→ CMNH

223 **Southern Broken-Dash**—♀—NC, Manteo—5·VIII·1972—S. M. Gifford →
    R. W. Surdick → CMNH

224 **Arogos Skipper**—♂—FL, Duval Co., Argyle Forest Blvd.—3·IX·1988—J. Slotten →
    F. Preston and J. Preston → MGCL

224 **Arogos Skipper**—♀—FL, Duval Co., Argyle Forest Blvd.—28·VIII·1988—J. Slotten →
    F. Preston and J. Preston → MGCL

225 **Delaware Skipper**—♂—PA, Monroe Co., Pocono Pines—9·VII·1992—D. M. Wright

225 **Delaware Skipper**—♀—PA, Monroe Co., State Game Lands 127—27·VII·1995—
    R. W. Surdick → CMNH

228 **Mulberry Wing**—♂—PA, Berks Co., Reading—8·VII·1959—J. Smaglinski → MGCL

228 **Mulberry Wing**—♀—PA, Schuylkill Co., Schuylkill Haven—7·VII·1969—W. Houtz →
    MGCL

229 **Zabulon Skipper**—♂—PA, Montgomery Co., Flourtown—ex ♀ 10·VIII·1987, em'g'd
    12·X·1987—R. W. Boscoe → MGCL

229 **Zabulon Skipper**—♀—PA, Montgomery Co., Flourtown—ex ♀ 10·VIII·1987, em'g'd
    29·X·1987—R. W. Boscoe → MGCL

230 **Hobomok Skipper**—♂—PA, Centre Co., Black Moshannon State Park—6·VI·1978—
    F. D. Fee → MGCL

230 **Hobomok Skipper**—♀—PA, Montgomery Co., Flourtown—ex ♀ 7·V·1988, em'g'd
    10·IX·1988—R. W. Boscoe → MGCL

231 **Hobomok Skipper**—♀—PA, Beaver Co., Raccoon Creek State Park—3·VI·1979—
    R. W. Surdick → CMNH

231 **Hobomok Skipper**—♀—PA, Centre Co., Black Moshannon State Park—ex ♀
    11·VI·1988, em'g'd 28·VIII·1988—R. W. Boscoe → MGCL

232 **Broad-winged Skipper** (ssp. *viator*)—♂—PA, Mercer Co., N. Liberty—22–31·VII·????
    —J Bauer → CMNH

232 **Broad-winged Skipper** (ssp. *viator*)—♀—PA, Mercer Co., N. Liberty—22–31·VII·????
    —J. Bauer → CMNH

233 **Broad-winged Skipper** (ssp. *zizaniae*)—♂—MD, Dorchester Co., Blackwater National
    Wildlife Refuge—ex ♀ 25·VI·1989, em'g'd 15·IX·1989—R. W. Boscoe → MGCL

233 **Broad-winged Skipper** (ssp. *zizaniae*)—♀—MD, Dorchester Co., Blackwater National
    Wildlife Refuge—ex ♀ 25·VI·1989, em'g'd 21·IX·1989—R. W. Boscoe → MGCL

233 **Broad-winged Skipper** (ssp. *zizaniae*)—♂—NC, Dare Co., East Lake—
    11·VIII·1967—R. W. Surdick → CMNH

234 **Black Dash**—♂—PA, Centre Co., Colyer Lake—9·VII·1975—MGCL

234 **Black Dash**—♀—PA, Huntingdon Co., Rte 26, W of Whipple Dam State Park—ex ♀
    19·VII·1995, em'g'd 27·VI·1996—R. W. Boscoe → MGCL

235 **Dion Skipper**—♂—PA, Tioga Co., Arnot—ex ♀ 16·VII·1988, em'g'd 8·X·1988—
    R. W. Boscoe → MGCL

235 **Dion Skipper**—♀—PA, Tioga Co., Arnot—ex ♀ 16·VII·1988, em'g'd 8·X·1988—
    R. W. Boscoe → MGCL

236 **Two-spotted Skipper**—♂—PA, Centre Co., Black Moshannon State Park—23·VI·1976
    —F. D. Fee → MGCL

236 **Two-spotted Skipper**—♀—PA, Clinton Co., Tamarack—21·VI·1975—F. D. Fee →
    MGCL

237 **Dun Skipper**—♂—PA, Huntingdon Co., Barnes Twp—28·VI·1978—F. D. Fee → MGCL

237 **Dun Skipper**—♀—PA, Montgomery Co., Flourtown—ex ♀ 28·VI·1990, em'g'd
2·IX·1990—R. W. Boscoe → MGCL

240 **Dusted Skipper**—♂—PA, Chester Co., Nottingham—27·V·1980—R. W. Boscoe →
MGCL

240 **Dusted Skipper**—♀—PA, Montgomery Co., Flourtown—21·V·1975—R. W. Boscoe →
MGCL

241 **Pepper and Salt Skipper**—♂—PA, Centre Co., Black Moshannon State Park—
11·VI·1978—F. D. Fee → MGCL

241 **Pepper and Salt Skipper**—♀—PA, Centre Co., Black Moshannon State Park—
1·VI·1978—F. D. Fee → MGCL

242 **Common Roadside Skipper**—♂—PA, Tioga Co., Stephenhouse Run, Middlebury
Twp—31·V·1969—G. Patterson → FEM

242 **Common Roadside Skipper**—♀—PA, Tioga Co., Lick Run, Gaines Twp—
20·VI·1967—G. Patterson → FEM

243 **Twin-spot Skipper**—♂—FL, Sarasota Co., Myakka River State Park—23·IV·1956—
D. L. Lindsley → MGCL

243 **Twin-spot Skipper**—♀—FL, Miami-Dade Co., Everglades NP—29·XII·1956—
D. L. Lindsley → MGCL

244 **Brazilian Skipper**—♂—FL, Okaloosa Co., Elgin AFB—29·VI·1962—H. O. Hilton →
MGCL

244 **Brazilian Skipper**—♀—FL, Orange Co., Ocoee—5·IX·1979—L. C. Dow → MGCL

245 **Salt Marsh Skipper**—♂—NJ, Cape May Co., Ocean View—ex ♀ 18·VI·1989, em'g'd
24·VII·1989—R. W. Boscoe → MGCL

245 **Salt Marsh Skipper**—♀—NJ, Cape May Co., Ocean View—ex ♀ 18·VI·1989, em'g'd
2·IX·1989—R. W. Boscoe → MGCL

246 **Ocola Skipper**—♂—FL, Pasco Co., New Port Richey—ex ♀ 26·X·1999, em'g'd
14·XII·1999—R. W. Boscoe → MGCL

246 **Ocola Skipper**—♀—FL, Pasco Co., New Port Richey—ex ♀ 26·X·1999, em'g'd
14·XI·1999—R. W. Boscoe → MGCL

247 **Mustard White**—Canada, Manitoba, The Pas—7·VII·1931—CMNH

247 **Mustard White**—Canada, Ontario, Nipigon, Orient Bay—?·VII·1914—C. K. Jennings →
CMNH

248 **Large Orange Sulphur**—♂—TX, Zapata Co., Falcon State Park—5·X·2200—MGCL

248 **Large Orange Sulphur**—♀—TX, Hidalgo Co., Sullivan City—16·X·1951—
W. J. Reinthal → MGCL

248 **Mexican Yellow**—♂—Mexico, Cuernavaca—27·VII·1927—MGCL

249 **Reakirt's Blue**—♂—Iowa, Polk Co., Des Moines—1·VIII·1925 → CMNH

249 **Reakirt's Blue**—♀—TX, Comal Co., New Braunfels, Landa Park—ex ♀ 29·IX·1986,
em'g'd 28·X·1986—R. W. Boscoe → MGCL

249 **Small Tortoiseshell**—Norway—pupa 15·VII·1985, em'g'd 26·VII·1985 → CMNH

250 **White Peacock**—FL, Miami-Dade Co., Homestead—6·XII·2012—J. L. Monroe →
MGCL

250 **Goatweed Leafwing**—♂—TX, Polk Co., Alabama and Coushatta Indian Res.—ex ♀
13·IV·1988, em'g'd 24·V·1988—R. W. Boscoe → MGCL

250 **Goatweed Leafwing**—♀—TX, Polk Co., Alabama and Coushatta Indian Res.—ex ♀ 18·III·1990, em'g'd 10·V·1990—R. W. Boscoe → MGCL

251 **Gemmed Satyr**—NC, Dare Co., East Lake—18·IV·1976—R. W. Surdick → CMNH

251 **Dukes' Skipper**—♂—VA, Norfolk Co., Indian Creek—S. S. Nicolay → MGCL

251 **Dukes' Skipper**—♀—VA, Virginia Beach, North Landing River.—ex ♀ 10·VI·1989, em'g'd 12·IX·1989—R. W. Boscoe → MGCL

252 **Eufala Skipper**—NC, Buxton, Cape Hatteras—13·VIII·1974—S. M. Gifford → R. W. Surdick → CMNH

Selected Faunal Lists
Pennsylvania

Anderson, Robert S. 1971. "Butterflies of the Serpentine Barrens of Pennsylvania." *Entomological News* 82: 5–12.

Barton, B. 1996. *Final Report on the Regal Fritillary 1992–1995: Fort Indiantown, Annville, Pennsylvania*. Unpublished report to US Department of Defense, National Guard. Annville, PA: Pennsylvania Science Office, Environmental Unit.

Boscoe, Richard W. 2012. "The Butterflies of Fort Washington State Park." Militia Hill Hawk Watch in association with Wyncote Audubon. http://wyncoteaudubon.org /wp-content/uploads/fwspleps.pdf.

Bramble, William C., Richard H. Yahner, and W. Richard Byrnes. 1997. "Effect of Maintenance with Treatments on Butterfly Populations of an Electric Transmission Right-of-Way in the Piedmont Region, Pennsylvania." *Journal of Arboriculture* 23: 196–206.

Bramble, William C., Richard H. Yahner, and W. Richard Byrnes. 1999. "Effect of Maintenance of an Electric Transmission Right-of-Way on Butterfly Populations." *Journal of Arboriculture* 25: 302–10.

Calhoun, John V., and David M. Wright. 2016. "Remarks on the Recent Publication of Titian R. Peale's "Lost Manuscript," Including New Information about Peale's Lepidoptera Illustrations." *Journal of Research on the Lepidoptera* 49: 21–51.

Clench, Harry Kendon. 1958a. "Common Spring Butterflies at Powdermill." *Carnegie Magazine* 32: 122–24.

Clench, Harry Kendon. 1958b. "The Butterflies of Powdermill Nature Reserve." Research Report No. 1, Powdermill Nature Reserve of the Carnegie Museum. In collection of Library of the Carnegie Museum of Natural History, Pittsburgh, PA.

Clench, Harry Kendon. 1959. "Powdermill Butterflies." Research Report No. 4, Powdermill Nature Reserve of the Carnegie Museum. In collection of Library of the Carnegie Museum of Natural History, Pittsburgh, PA.

Clench, Harry Kendon. 1960. "The Butterflies of Powdermill Nature Reserve." Research Report No. 5, Powdermill Nature Reserve of the Carnegie Museum. In collection of Library of the Carnegie Museum of Natural History, Pittsburgh, PA.

Clench, Harry Kendon. 1968. "Revised List of Powdermill Butterflies." Research Report No. 21, Powdermill Nature Reserve of the Carnegie Museum. In collection of Library of the Carnegie Museum of Natural History, Pittsburgh, PA.

Davis, R. N. 1915. *Illustrated and Annotated Catalog of the Butterflies of Lackawanna County, Pennsylvania*. Everhart Museum Natural History Bulletin No. 1. Scranton, PA: Everhart Museum of Natural History, Science and Art.

Ehle, G. E. 1981. *Checklist of Butterflies of Lancaster County, Pennsylvania*. Mimeograph, distributed personally by author. North Museum, Lancaster, PA.

Engel, Henry. 1908. "A Preliminary List of the Lepidoptera of Western Pennsylvania Collected in the Vicinity of Pittsburgh." *Annals of the Carnegie Museum* 5 (1): 27–136.

Ferster, Betty, Betsy Ray Leppo, Mark T. Swartz, Kevina Vulinec, Fred Habegger, and Andrew Mehring. 2008. "Lepidoptera of Fort Indiantown Gap National Guard Training Center, Annville, Pennsylvania." *Northeastern Naturalist* 15 (1): 141–48.

Genoways, Hugh H., and Fred J. Brenner, eds. 1985. *Species of Special Concern in Pennsylvania*. Carnegie Museum of Natural History Special Publication No. 11. Pittsburgh, PA: Carnegie Museum of Natural History.

Keller, Gregory S., and Richard H. Yahner. 2002. "Butterfly Communities in Two Pennsylvania National Parks." *Northeastern Naturalist* 9 (2): 235–42.

Klinger, Mark A., John E. Rawlins, Charles W. Bier, Ted Walke, and Wild Resource Conservation Fund. 1993. Poster, *Butterflies and Skippers of Pennsylvania*. Harrisburg, PA: Wild Resource Conservation Fund.

McWilliams, Gerald M. 1982. *Checklist of Butterflies Recorded in Erie County, PA, Including Presque Isle State Park*. Erie, PA: Presque Isle Audubon Society.

Mumbauer, J. 1994. "Butterflies. (Lepidoptera)." In *Flora and Fauna in the Perkiomen Valley*, 6–7. Bedminster, PA: Adams Apple Press. First published 1922.

Patterson, George F. 1971. "Unusual Butterflies in Northern Pennsylvania." *Journal of the Lepidopterists' Society* 25 (3): 222.

Prescott, J. M. 1984. "The Butterflies of Presque Isle State Park, Erie County, Pennsylvania." *Melsheimer Entomological Series* 34: 19–23.

Quinter, E. L. 1971. *A Preliminary List of the Papilionoidea and Hesperioidea of Schuylkill County, Pennsylvania*. Mimeograph, distributed personally by the author. In possession of the author.

Rawlins, John E., and Charles W. Bier. 1998. "Invertebrates: Review of Status in Pennsylvania." In *Inventory and Monitoring of Biotic Resources in Pennsylvania*, ed. J. D. Hassinger, R. J. Hill, G. L. Storm, and R. H. Yahner, 85–120. Current Ecological and Landscape Topics 1. University Park, PA: Pennsylvania Biological Survey, Center for Biodiversity Research, Pennsylvania State University.

Rawlins, John E., with Charles W. Bier and Betsy Ray Leppo, contributing primary investigators. 2007. *Pennsylvania Comprehensive Wildlife Conservation Strategy. Invertebrates*. Version 1.1. Report submitted to the Pennsylvania Game Commission and Pennsylvania Fish and Boat Commission, January 12. http://fishandboat.com/promo/grants/swg/nongame_plan/pa_wap_sections/appx05_invertebrates.pdf.

Shapiro, Arthur M. 1963a. "The Butterflies of Morris Arboretum." *Morris Arboretum Bulletin* 14: 8–14, 32–36.

Shapiro, Arthur M. 1963b. "The Butterflies of Morris Arboretum: 1963." *Morris Arboretum Bulletin* 14: 67–69.

Shapiro, Arthur M. 1966. *Butterflies of the Delaware Valley*. Philadelphia, PA: American Entomological Society.

Shapiro, Arthur M. 1968. First Additions to "Butterflies of the Delaware Valley." *Entomological News* 79: 150–52.

Shapiro, Arthur M. 1970. "The Butterflies of the Tinicum Region." In *Two Studies of Tinicum Marsh, Delaware and Philadelphia Counties, Pennsylvania*, 95–104. Washington, DC: Conservation Foundation.

Skinner, Henry, and E. M. Aaron. 1889a. "A List of the Butterflies of Philadelphia, Pa." *Canadian Entomologist* 21 (7): 126–31.

Skinner, Henry, and E. M. Aaron. 1889b. "A List of the Butterflies of Philadelphia, Pa." *Canadian Entomologist* 21 (8): 145–49.

Stamm, Rev. J. C. 1911. "Butterflies of Montour County, Pennsylvania (Lepid.)." *Entomological News* 22: 422–23.

Tietz, Harrison Morton. 1952. *The Lepidoptera of Pennsylvania: A Manual.* State College, PA: Pennsylvania State College, School of Agriculture, Agriculture Experiment Station.

Williams, R. C., Jr. 1941. "A List of Butterflies which May Be Found within 50 Miles of Philadelphia (Lepid.: Rhopalocera)." *Entomological News* 52 (8): 217–19.

Wright, David M. 2015. *Atlas of Pennsylvania Butterflies.* 14th ed. Lansdale, PA: distributed personally by the author.

Yahner, Richard H. 1996. "Butterfly and Skipper Communities in a Managed Forested Landscapes." *Northeast Wildlife* 53: 1–9.

Yahner, Richard H. 1997a. *Biodiversity Conservation of Butterflies and Skippers in Residential Landscapes of Pennsylvania. Final Report.* Harrisburg, PA: Wild Resource Conservation Fund.

Yahner, Richard H. 1997b. "Butterfly and Skipper Use of Nectar Sources in Forested and Agricultural Landscapes of Pennsylvania." *Journal of the Pennsylvania Academy of Sciences* 71: 104–7.

Yahner, Richard H. 1999. "Edge Use by Butterfly Communities in Agricultural Landscapes." *Northeast Wildlife* 54: 13–24.

Yahner, Richard H. 2001. "Butterfly Communities in Residential Landscapes in Central Pennsylvania." *Northeastern Naturalist* 8 (1): 113–18.

Adjacent States

Allen, Thomas J. 1997. *The Butterflies of West Virginia and Their Caterpillars.* Pittsburgh, PA: University of Pittsburgh Press.

Gochfeld, Michael, and Joanna Burger. 1997. *Butterflies of New Jersey: A Guide to Their Status, Distribution, Conservation, and Appreciation.* New Brunswick, NJ: Rutgers University Press.

Iftner, David C., John A. Shuey, and John V. Calhoun. 1992. *Butterflies and Skippers of Ohio.* Bulletin of the Ohio Biological Survey, n.s. 9. Columbus, OH: College of Biological Sciences, Ohio State University.

Shapiro, Arthur M. 1974. *Butterflies and Skippers of New York State.* Ithaca, NY: Cornell University.

Woodbury, Elton N. 1994. *Butterflies of Delmarva.* Centreville, MD: Tidewater Publishers and Delaware Nature Society.

Eastern North America

Cech, Rick, and Guy Tudor. 2005. *Butterflies of the East Coast: An Observer's Guide.* Princeton, NJ: Princeton University Press.

Glassberg, Jeffrey. 1993. *Butterflies through Binoculars: A Field Guide to Butterflies in the Boston–New York–Washington Region.* New York: Oxford University Press.

Glassberg, Jeffrey. 1999. *Butterflies through Binoculars: The East.* New York: Oxford University Press.

Klots, Alexander B. 1951. *A Field Guide to the Butterflies of North America, East of the Great Plains.* Peterson Field Guide Series. Boston, MA: Houghton Mifflin.

Layberry, Ross A., Peter W. Hall, and J. Donald Lafontaine. 1998. *The Butterflies of Canada.* Toronto, ON: University of Toronto Press.

Minno, Marc C., and Maria Minno. 1999. *Florida Butterfly Gardening: A Complete Guide to Attracting, Identifying, and Enjoying Butterflies of the Lower South*. Gainesville: University Press of Florida.

Opler, Paul A., and George O. Krizek. 1984. *Butterflies East of the Great Plains: An Illustrated Natural History*. Baltimore, MD: Johns Hopkins University Press.

Opler, Paul A. and Vichai Malikul. 1992. *A Field Guide Guide to Eastern Butterflies*. Peterson Field Guides. Boston, MA: Houghton Mifflin.

Butterfly Gardening

Mank, Judith, and Margaret C. Brittingham. 2016. "Gardening for Butterflies." Pennsylvania Wildlife No. 8. University Park, PA: Agricultural Communications and Marketing, Pennsylvania State University. http: //extension.psu.edu/natural-resources/wildlife /landscaping-for-wildlife/pa-wildlife-8/extension_publication_file.

North American Butterfly Association. 2016. "Basics of Butterfly Gardening." *Butterfly Garden and Habitat Program*. Accessed August 25. http://nababutterfly.com /start-butterfly-garden/.

Richael, Ron. 2010. *Attracting Butterflies: A Handbook for Butterfly Gardening*. Bloomington, IN: Xlibris.

Sutton, Patricia. 2014a. "How to Create a Butterfly and Hummingbird Garden, Part 1: Know Who You Are Inviting." *New Jersey Audubon*. http://www.njaudubon.org /SectionBackyardHabitat/CreateaGarden.aspx.

Sutton, Patricia. 2014b. "How to Create a Butterfly and Hummingbird Garden, Part 2: Planning the Garden." *New Jersey Audubon*. http://www.njaudubon.org /SectionBackyardHabitat/HowtoCreateAButterflyandHummingbirdGarden.aspx.

Caterpillar Identification

Allen, Thomas J., James P. Brock, and Jeffrey Glassberg. 2005. *Caterpillars in the Field and Garden: A Field Guide to the Butterfly Caterpillars of North America*. New York: Oxford University Press.

Wagner, David L. 2005. *Caterpillars of Eastern North America: A Guide to Identification and Natural History*. Princeton Field Guides. Princeton, NJ: Princeton University Press.

Physiographic Province Map

US Geological Survey and Pennsylvania Geological Survey, compilers. 2008. Digital shaded-relief map of Pennsylvania, (tabloid ed.): Pennsylvania Geological Survey, 4th ser., Map 65, scale 1:1,350,000. Available online as a ZIP file. http://www.dcnr.state.pa.us /topogeo/publications/pgspub/template/index.htm?id=742.

State Distribution Maps and Flight Phenograms

Butterflies and Moths of North America. 2016. Accessed September 12. http://www .butterfliesandmoths.org/.

Field Summary of The Lepidopterists' Society (*Lepidopterists' News*, 1947–1953).

Season Summary of The Lepidopterists' Society ("News of the Lepidopterists' Society," 1959–2014).

North American Butterfly Association. 2012. *Sighting Archive, 2000–2011*. Last updated March 22. http://www.naba.org/sightings/Archives/SightingsArchives.htm.

North American Butterfly Association. 2016. "Recent Sightings." *NABA Sightings*. https://sightings.naba.org/.

PA Leps and Odes mailing list. 2005–2016. Message history available at http://groups.yahoo.com/group/PaLepsOdes/.

Scientific Names

Pelham, Jonathan P. 2012. *Catalogue of the Butterflies of the United States and Canada*. Last revised February 14. http://butterfliesofamerica.com/US-Can-Cat-1-30-2011.htm.

Common Names

North American Butterfly Association. 2001. *Checklist and English Names of North American Butterflies*. 2nd ed. Morristown, NJ: North American Butterfly Association.

North American Butterfly Association. 2016. *Checklist of North American Butterflies Occurring North of Mexico—Edition 2.3*. Last revised May 3. http://www.naba.org/pubs/enames2_3.html.

Lepidopterist Organizations

The Lepidopterists' Society. http://www.lepsoc.org/. Journal, newsletter, season summary, annual meeting.

North American Butterfly Association. http://www.naba.org/. Journal, butterfly gardening, butterfly counts, recent sightings, local chapters.

Xerces Society. http://www.xerces.org/. Conservation of invertebrates, journal, fact sheets.

Museum Collections

Academy of Natural Sciences of Philadelphia, Philadelphia, PA
Allyn Museum of Entomology, Sarasota, FL
American Museum of Natural History, New York, NY
Boston University—Boston Society of Natural History, Boston, MA
Butler University, Department of Biological Sciences, Indianapolis, IN
California Academy of Sciences, San Francisco, CA
Carnegie Museum of Natural History, Pittsburgh, PA
Cleveland Museum of Natural History, Cleveland, OH
Cornell University Insect Collection, Ithaca, NY
Dayton Museum of Natural History, Dayton, OH
Delaware County Institute of Science, Media, PA
Eastern College, University, Biology Department, Saint Davids, PA
Everhart Museum, Scranton, PA
Field Museum of Natural History, Chicago, IL
Florida State Collection of Arthropods, Gainesville, FL
Frost Entomological Museum, State College, PA
Illinois Natural History Survey, Champaign, IL

Indiana State Museum, Indianapolis, IN
Los Angeles County Museum of Natural History, Los Angeles, CA
McGuire Center for Lepidoptera and Diversity, Gainesville, FL
Michigan State University, Department of Entomology, East Lansing, MI
Museum of Comparative Zoology, Cambridge, MA
Natural History Museum—London, London, England
National Museum of Natural History (Smithsonian), Washington, D.C.
New Jersey State Museum, Trenton, NJ
Newark Museum, Newark, NJ
North Museum of Nature and Science, Lancaster, PA
Oakes Museum of Natural History, Mechanicsburg, PA
Ohio State Museum of Biological Diversity, Columbus, OH
Pennsylvania Department of Agriculture, Harrisburg, PA
Pennsylvania Department of Forestry, Middletown, PA
Pennsylvania Natural Diversity Inventory, Middletown, PA
Pennsylvania State Museum, Harrisburg, PA
Purdue University, Department of Entomology, West Lafayette, IN
Reading Museum, Reading, PA
Royal British Columbia Museum, Victoria, British Columbia
Royal Ontario Museum, Toronto, Ontario
Rutgers University, Department of Entomology, New Brunswick, NJ
Staten Island Institute of Arts and Sciences, Staten Island, NY
Tioga Point Museum, Athens, PA
University of California, Davis, Bohart Museum, Davis, CA
University of California, Riverside, Department of Entomology, Riverside, CA
University of Connecticut, Department of Ecology and Evolutionary Biology, Storrs, CT
University of Delaware, Department of Entomology and Wildlife Ecology, Newark, DE
University of Guelph, Insect Collection, Guelph, Ontario, Canada
University of Louisville, Department of Biology, Louisville, KY
University of Michigan Museum of Zoology, Ann Arbor, MI
University of Pennsylvania, Department of Biology, Philadelphia, PA
Wagner Free Institute of Science, Philadelphia, PA
Western Illinois University, Department of Biological Sciences, Macomb, IL
Yale Peabody Museum of Natural History, New Haven, CT

Butterfly and Skipper Common and Scientific Names

clovers: 59, 93, 177
clover
    white, 58
    red: 58
    golden: 190
cockscomb, feather: 197
cohosh, black: 98
Collinsonia canadensis: 96
columbine, wild: 189
coneflower, cutleaf: 131
Cornus
    alternifolia: 97
    amomum: 96
    florida: 94
    racemosa: 96
    sericea: 97
Coronilla varia: 93, 190
Corylus cornuta: 91
cranberry: 73
Crataegus sp.: 80
crucifers: 52
currants: 145
Cynodon dactylon: 206
Dactylis glomerata: 167, 206, 214, 216
Danthonia spicata: 208, 212
deer tongue: 216, 222, 225, 230
Desmodium sp.: 89, 93, 172, 177–78, 179
Desmodium
    canadense: 175
    paniculatum: 175
    rotundifolium: 175
Dichanthelium
    acuminatum: 222, 225
    clandestinum: 216, 222, 225, 230
Digitaria sp.: 204, 206
dill: 34
Distichlis spicata: 245
dock: 72
Doellingeria umbellata: 131–32
dogwood
    alternate-leaf: 97
    flowering: 94
    red osier: 97

Dutchman's pipe: 30
Echinops sp.: 153
Eleusine indica: 206
elm: 141–42, 148, 150
Epigaea repens: 83
Eragrostis sp.: 208, 229
Erianthus sp.: 225
everlasting, pearly: 152
Fagus grandifolia: 91
fescue: 168
Festuca sp.: 168
foxglove, yellow false: 133
Fraxinus americana: 38
gerardia, purple: 138
galls: 97
goatsbeard: 99
grass
    bentgrass: 242
    Bermuda: 206
    blue: 219, 230
    blue joint: 198, 229
    bottlebrush: 163
    crab: 204, 206
    goose: 206
    Kentucky bluegrass: 242
    lovegrass: 208, 229
    oat: 242
    orchard: 167, 206, 214, 216
    panic: 222, 225
    plume: 225
    poverty oat: 208, 212
    reed canary: 202, 225
    rice cutgrass: 202, 246
    salt: 245
    switch: 208, 210, 225
    timothy: 203
hackberries: 111, 141–42
hackberry
    common: 160–61
    dwarf: 160–61
hawthorns: 80
hazelnut, beaked: 91
Helianthus tuberosus: 131
hickories: 79
hickory
    bitternut: 79
    pignut: 78
hog peanut: 96, 172, 177

hollyhock: 196
hop, Japanese: 141–42, 150
hops, hop vine: 111
hoptree: 36, 38
horsebalm: 96
Humulus
    japonicus: 141–42, 150
    lupulus: 111
Hystrix patula: 163
indigo, wild: 84, 172, 187, 190–91
indigo bush, false: 62, 172
ironweed, New York: 131
Juglans
    cinerea: 78
    nigra: 78
Juniperus virginiana: 87
Kalmia latifolia: 82
lamb's quarters: 194, 197
Laportea sp.: 154
laurel, mountain: 82
Leersia oryzoides: 202, 246
legumes: 58–59, 89, 92–93, 172–73, 177–79, 188
Lespedeza sp.: 89, 93, 172, 177–78, 187
Lindera benzoin: 46
Liriodendron tulipifera: 38
locust, black: 172, 182, 187
loosestrife: 89
lousewort, Canadian: 133
lupine, wild: 84, 103, 191
Lupinus perennis: 84, 103, 191
Lythrum sp.: 89
mallow
    common: 153, 196
    marsh: 196
    round-leaved: 196
Malva
    neglecta: 153, 196
    rotundifolia: 196
meadowsweet, white: 96
Medicago
    lupulina: 177
    sativa: 92–93
medick, black: 177

Hostplant Common and Scientific Names

Medium

White bands, orange bands, eyes,
Ladies, Red Admiral, and Buckeye pp. 138, 139, 152–154

Grays, browns and tan with eyespots,
Emperors, Satyrs, and Ringlet, pp. 160–169

Medium to small

Brown with bands and large spots, Dicot and Spread-wing Skippers, pp. 172–179

Predominantly brown with small white spots, Duskywings, pp. 182–191

Predominantly orange and black with intricate patterns,
Checkerspots and Crescents, pgs. 130–137